THE WISDOM OF THE EYE

THE WISDOM
OF THE EYE

David Miller

ACADEMIC PRESS

San Diego   London   Boston   New York   Sydney   Tokyo   Toronto

Academic Press
A Harcourt Science and Technology Company
525 B Street, Suite 1900, San Diego, California 92101-4495, USA
http://www.academicpress.com

Academic Press
24-28 Oval Road, London NW1 7DX
http://www.hbuk.co.uk/ap/

Library of Congress Card Number: 00-100272

International Standard Book Number: 0-12-496860-0

PRINTED IN THE UNITED STATES OF AMERICA
00 01 02 03 04 05 MM 9 8 7 6 5 4 3 2 1

CONTENTS

INTRODUCTION VII
PREFACE IX

PART I

THE EYE

1

THE YOUNG EYE 3

2

THE IMAGE OF THE ADULT HUMAN EYE 27

3

EYES OF DIFFERENT ANIMALS 45

4

THE HEALING EYE 61

5

REFRACTIVE ERRORS OF THE HUMAN EYE: A SOCIOLOGIC VIEWPOINT   73

6

EYE COMMUNICATION   85

PART II

THE VISUAL BRAIN

7

CREATING VISUAL STORIES AND ILLUSIONS AROUND THE RETINAL IMAGE   103

8

BRAIN SHARPENING OF THE RETINAL IMAGE   135

9

COLORING THE RETINAL IMAGE   147

10

AWARENESS OF THE RETINAL IMAGE   161

INDEX   165

INTRODUCTION

The knowledge explosion in the field of basic eye and vision research has been enormous. Unhappily, the acquisition and understanding of much of this new basic information is beyond the grasp of most eye clinicians, students, and scientists from other fields. Aside from simply the large amount of material, there are other reasons for the dilemma.

As a field matures its language becomes more rigorous or technical. Thus, in order for a scientist's work to be accepted by his or her peers it must be written in the most current technical terms, which (unhappily) makes it difficult to follow for the rest of us. Therefore, this book uses simple language and anecdotes whenever possible to present the information. Admittedly, a few details may be lost in translation.

A second problem is the current practice of reporting data with a minimum of interpretation. Therefore, in this book information is placed into workable concept packages. Of course, theories and concepts will not always account for 100% of the published data. However, this approach of emphasizing conceptual packages not only helps the reader in organizing the data but is very much in keeping with the way the visual brain works. In the chapter on the brain processing of the retinal image, the visual brain is depicted as almost always developing a conceptual hunch about a scene when the visual information is incomplete. For example, the visual brain creates a complete picture when key objects are occluded, when retinal images of objects fall into physiologic blind spots or scotomas, or when important scene changes do not register because they occur during blinks, flickers, and saccades.

A third device is used in this book to make the material more understandable. A very broad theme is used to tie all of the chapters together. This theme suggests that "wisdom of the eye and visual brain" significantly helped primitive humans to survive in their early hostile environment.

But how relevant are the visual survival strategies of the primitive hunter/gatherer to us? I would maintain that we are those people in modern dress. Recall that the appearance of the first human is said to have occurred about 1.8 million years ago. As this creature evolved it remained a hunter/gatherer until farming got started some 10,000 years ago. Thus, one could say that we humans have been hunter/gatherers for over 99% of our existence.

At this point, one might question the thesis that vision was a key survival device for the early human since aspects of our eye and vision are inferior to that of other creatures. After all, the eagle eye has a higher level of resolution than our eye. The injured rabbit, newt, and sculpin eyes heal more efficiently than ours, the butterfly eye sees a broader range of wavelengths than we do, and most mammals with eyes on the side of their head have a wider field of view than ours. Indeed, our system must be appreciated as functioning reasonably well under a wide variety of different circumstances. For example, our optical system can resolve a separation between two objects of one minute of arc (i.e., separation of two people 10 feet apart, from a distance of over 3,000 feet away). Our visual brain does highlight features of the retinal image that imply danger, and our eyes can repair most common surface injuries within a few days. On the other hand, we are unique in transmitting a wide vocabulary of emotions and subconscious thoughts through our eyes.

PREFACE

This book may be seen as a survey of the major concepts underlying many of the findings of the basic sciences related to the human eye and visual brain. Secondarily, the book is an attempt to tie all of the subjects (i.e., chapters) together with a very broad theme. The broader theme suggests that the "wisdom" referred to in the book title are the elements of the eye and visual brain that have significantly helped early human societies survive.

In Chapter 1, the anatomy and physiology of the infant eye are discussed in terms of the visual level needed to survive. The infant's small-sized eye and undeveloped neural circuitry do not provide the visual acuity necessary to distinguish items as small as the lettering in the telephone book. However, the baby is able to read the expressions on its mother's face when she is near, follow her eye movements, and sweep away an annoying insect. The last section of the chapter discusses the drawings of children. This section discusses how an individual culture can influence the way children are taught to see objects and events.

Chapter 2, on the human adult retinal image, is really a discussion of the natural optical elements that create the human adult retinal image. The reasons behind the limits of human resolution are described as well as the way that nature copes with the optical aberrations. The chapter closes by introducing the idea that the human adult eye has a number of functions. Optically it must produce a respectable level of resolution, a large range of focus, and a visual capacity in dim light. On the other hand, the eye must contain the equivalent of a first aid station and must also be capable of communicating deep-felt emotion. Therefore, the adult human eye is seen as a multifunctional organ that strives to attain the proper balance and compromise for each function.

Chapter 3 on animal vision focuses on the great "visual athletes" of the animal kingdom. Specifically, the section concerning the eagle eye describes its uncanny ability to visually perform in a very glaring situation as well as resolve objects at least one quarter the size that we can. The diving bird, the hooded merganser, has an accommodation range eight times that of a 20-year-old human. The cat, with its special reflecting retroretinal membrane, can sense a light far dimmer than we can. By looking closely at these and other "visual athletes," the chapter tries to define the physical limits of animal visual function and relate these visual abilities to survival in natural niches. The reader is now better able to compare the level of optical function of the human eye with the examples seen within nature's spectrum.

Chapter 4 concerns eye injuries. It tackles the puzzling finding that in the underdeveloped world, the incidence of serious eye injuries is quite high, yet the incidence of blindness from such injuries accounts for only 1% to 2% of all blindness. In presenting five different injury cases of escalating severity, the reader gets to appreciate the short-term and long-term survival strategies used by the body as it naturally heals an eye wound.

Chapter 5 discusses refractive errors. It describes the optics and epidemiology of each type of error. The approach to this subject is made a bit unconventional by discussing the possible advantages of each type of refractive error. For example, in the ancient Orient, high myopes were respected because of the ability to make fine embroidery or carve tiny Netsuke statues. On the other hand, lookouts in primitive tribes with "with the rule" astigmatism could see a far-off vertical figure better than a normal-sighted person. The chapter closes with the thought that there might be advantages in having a spectrum of people with different refractive errors in a primitive community (i.e., prior to the use of spectacles).

Chapter 6 describes the eye and its surrounding tissue as a transmitter rather than a receiver of information. Indeed the same structures that nourish and protect the eye (lids, brows, conjunctiva, lacrimal gland) also facilitate communication of emotion between communal members. Specifically, it is our crying, our menacing brows, our locking of glances, our pupillary dilation and so on that convey our deepest emotions and help bind a social group together in order to strengthen its interdependence. The second half of the chapter moves into a discussion of animal eye spots (replicas of the external eye). These are found on the bodies of fish, wings of moths and butterflies, and even on the buttocks of a species of South American frogs, all complete with a central white dot representing the corneal reflex. Clearly, these spots suggest the presence of a larger animal and function as effective survival tools themselves.

The four chapters of Part II focus on the function of the visual brain. Chapter 7 is primarily concerned with optical illusions. These illusions (really the brain's distortion of the retinal image) probably had value for our primitive ancestors. Recall that the modern human (Cro-Magnon human) had its origin 100,000 years ago. For 90% of this period, we were hunter/gatherers. In seeking utility for these illusions, we must appreciate them through "the eyes" of these ancestors. Therefore, an attempt is made to reconstruct natural real-life replicas of these illusions when possible and then speculate as to how they might have worked to enhance the survival of the early ancestors. For example, certain illusions may have compensated for the relative slowness of neural processing time and helped a hunter hit a running animal. Other illusions alert the beholder to erect vertical objects, which might represent an attacker. Many of the illusions allow us to recognize human faces even when many details are obscured. These latter illusions clearly maintained the social contact needed to strengthen communal bonding.

Chapter 8 describes the ways in which the brain sharpens the most relevant features of the retinal image. Just as the computer enhances the images of the galaxies and planets beyond the optical resolution of our celestial telescopes, this chapter illustrates the manner in which the brain enhances the edges, the contrast, and the details of the retinal image, as well as filling in information when part of an object is covered. The last section of the chapter discusses the aging of the vision processing system. Evidence is presented that shows that although an older person may have a visual acuity of 20/20, their visual complaints could be related to slower brain processing of visual information or a poorer ability to cope with visual distractions, such as peripheral objects jumping into view while keeping one's eye on the road ahead.

Chapter 9 deals with human color vision. The chapter presents short biographical sketches of Newton, Young, Maxwell, and Land and their theories of color vision. What follows is a description of the many ways that the brain strengthens the impact of the colored retinal image. For example, the function of color constancy ensures that an object will have the same color appearance whether it is illuminated by sunlight at dawn, noon, or late afternoon. In another vein, the enhancement of color against certain backgrounds allows us to recognize more readily the colored patterns of poisonous insects and snakes. Color also makes our dreams more vivid and memories of events more long lasting. The chapter closes with some speculation about how a large meteorite crashing into the earth some 65 million years ago may have influenced the evolution of primate color vision.

Chapter 10 tackles the little discussed phenomenon of visual awareness, that is, the ability to be consciously aware that you are looking at something or someone. The chimpanzee brought up in a social environment (versus isolation) is the only primate besides the human that can recognize itself in the mirror. Interestingly, many autistic patients report that when looking in the mirror, they see, not themselves, but someone who always accompanies them. What then follows are descriptions of brain-injured patients who claim to see but really cannot, and others who claim not to see but can (blind sight). The chapter closes by speculating on the advantages of being aware of what we see.

THE EYE

1

THE YOUNG EYE

1. Relevant Early Anatomy
   A. *Axial Length*
   B. *Retinal Receptors*
2. Relevant Early Physiology
3. Recognizing Faces
4. Line Orientation Receptors
5. Monitoring Other's Eye Movements
6. Recognizing Movement
7. Recognizing Three Dimensions
8. Blockage of Visual Development (Amblyopia)
9. What Child Art Teaches Us About Visual Development
   A. *Drawing Faces*
   B. *The Influence of Culture on Child Art*
Summary
   A. *Social Seeing*
   B. *Amblyopia*
   C. *Child Art*

Primate and human infants must normally pass head first through their mother's pelvis in order to accommodate the limited opening determined by the bony configuration. Therefore, the size of the mother's pelvis limits the head and brain size of the infant. Specifically, the infant brain size of an ape is 55% of its full size, and the modern human infant brain is only 23% of the adult size.[1] The result is that human infants are neurologically very immature.[2] The baby monkey can immediately cling tightly to the fur on its mother's stomach, whereas the human infant has poor muscle strength and little motor control and is completely dependent on its mother for survival. During its immature period, the human infant lives in a very restricted and artificial reality, interacting primarily with its mother. It interacts very little with the forces of life in the outside world.

I would like to suggest that nature takes positive advantage of this early immaturity and restricted world contact. The infant's restricted curriculum concentrates on a few priorities in order to survive. Without words, the infant must be able to announce all its needs as well as encourage a high level of motherly devotion. To communicate with its mother, it must be able to read facial expressions and respond with a nonverbal vocabulary. What vision equipment does the infant have to perform these functions?

1. RELEVANT EARLY ANATOMY

A. AXIAL LENGTH

I first came to appreciate the small size of a baby's eye as a volunteer ophthalmologist in Tunisia. I needed a donor cornea for a patient who required a corneal transplant. Unhappily, the Eye Banking system was poorly organized and freshly donated eyes were not easily available. However, there was another source of corneas. The eyes of stillborn infants could be the answer.* Tunisian parents rarely took a dead baby home from the hospital. Why not pass on the gift of sight to those with corneal blindness? It was certainly a real life experiment. With the help of one of the Tunisian ophthalmologists, I was quietly ushered into the nursery of the obstetrical hospital one dark night. Opening the door to the nursery was like snapping on a radio and instantly turning the volume way up, as the babies all seemed to wail in unison. The nursery reminded me of a supermarket, with three rows of shelves circling the room. Each shelf was divided into compartments large enough to hold one baby. The babies themselves were all snugly wrapped, with their heads facing the center of the room. The little packages hardly stirred in their tight wrappings. Between some of the screamers, several babies slept peacefully. Then the head nurse led me into another room. The baby package on the table looked exactly like the other packages in the nursery, except this one neither cried nor moved. This was a stillborn, whose mother had already left the hospital. The stillborn corneas were slightly smaller and thinner than in the adult. However, they seemed suitable for transplantation. As I started to surgically remove both eyes in the nursery, I realized how much smaller the eyes were than those of the adult.

Larsen[3] noted that the axial length of the eye of the neonate was 17 mm and increased 25% by the time of adolescence. Note in Figure 1.1 that the size of the normal infant's eye is about three-fourths of the adult size. Geometric optics teaches us that the retinal image of the normal infant eye will therefore be about three-fourths the size of the adult's image.† A smaller image also means that much less fine detail is recorded. Figure 1.2 attempts to illustrate this principle by

---

* Newborn corneas are a bit more steeply curved and more elastic than adult corneas. Thus, corneal transplant surgeons prefer adult donor corneas if given a choice.

†Specifically, the size of the retinal image is dependent on an entity known as the nodal distance, which is 11.7 mm in the newborn and 16.7 mm in the adult, giving a ratio of adult to infant retinal images of 1.43. (Banks MS, Bennett PJ: Optical and photoreceptor immaturity limit the spatial and chromatic vision of human neonates. *J. Opt. Soc. Am. (A)* 5:2059, 1988.)

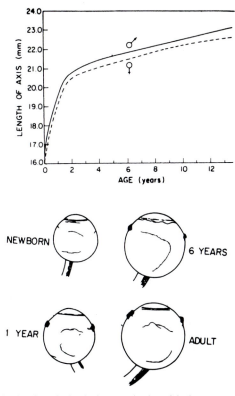

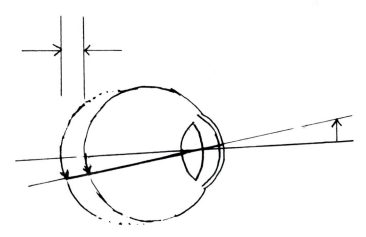

FIGURE 1.1 The drawings depict the increase in size of the human eye with age. (From Larsen SJ: Sagittal growth of the eye. II. Ultrasonic measures of the axial length of the eye from birth to puberty. *Acta Ophthalmol. (Copenh)* 49:837, 1971, with permission.)

FIGURE 1.2 The telephoto lens increases the size of the image, so that more detail can be seen. A longer eye accomplishes the same effect.

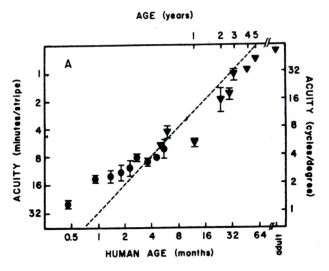

FIGURE 1.3   A graph showing the improvement of visual acuity in the baby as it ages. The method of preferential viewing was used to achieve these results. (From Teller DG: The development of visual function in infants. In *Vision and the Brain* (Cohen B, Bedis-Wollner I, eds.). Raven Press, New York, 1990, with permission.)

comparing a smaller regular camera image with one produced by a camera with a telephoto lens. The small retinal image may be but one reason why infant visual acuity is poorer than that of the adult. In fact, experiments have shown that the neonate's visual appreciation for fine detail at birth is $1/30$ or about 3% of the level of the adult*, yet it appreciates large objects (nose, mouth, eyes of close faces) as does the adult. The graph pictured in Figure 1.3 shows that visual acuity swiftly improves so that by the age of 12 months, the infant's level of visual acuity is 25% (20/80) of optimal adult visual acuity. This improvement in acuity seems to parallel eyeball growth. 20/20 vision will usually be reached by the age of 5.[1,4–7]

What other factors beyond eye size account for the young child's lowered visual acuity? As the eye grows, the optical power of the eye lens and cornea must weaken in a tightly coordinated fashion, so that the world stays in sharp focus on the retina. If this tight coordination of growth fails, the infant may become nearsighted or farsighted. This will be discussed further in another chapter. However, this brings up an interesting question. Given the fact that the coordination of eye length growth, and focusing power of the cornea and eye lens may be imperfect, is there some compensation provided, early in life, guaranteeing that almost every child can get a sharply focused retinal image of the world? Accommodation is the

---

* The infant's visual acuity is about 20/600 versus the normal visual acuity of an adult of 20/20. Infants, however, have enormous focusing capacity. Alas, we progressively lose accommodation as we age. More on this in Chapter 5.

safety valve that can help provide a sharp image, even if all the ocular components are not perfectly matched. In the young child, the range of accommodation is over 20 diopters. This, along with the fact that almost all infants eyes are far-sighted, means that most young eyes can get almost any object in focus by using part or all of this enormous focusing capacity.

A second factor that helps the infant achieve a sharper retinal image is an increased depth of focus, because of its smaller pupil.[6] Photographers use this device when they use larger F stops (F32, F64), to keep objects at different distances all in focus.

a. Emmetropization

The coordination of the power of the cornea, eye lens, and axial length in order to get a sharp retinal image of a distant object is known as *emmetropization.* In the United States over 70% of the population is either emmetropic or mildly hyperopic (easily corrected with a small accommodative effort).

Let's take a closer look at the coordinated changes with age of the cornea, lens, and axial length. In a nutshell, the optical components (cornea and lens) must lose refractive power as the axial length increases, so that a sharp image remains focused on the retina.

The cornea, which averages 48 D of power at birth and has an increased elasticity, will lose about 4 D by age 2.[8,9] One may assume that the spurt in growth of the sagittal diameter of the globe during this period pulls the cornea into a flatter curvature. The fact that the average corneal diameter is 8.5 mm at 34 weeks gestation, 9 mm at 36 weeks, 9.5 mm at term, and about 11 mm in the adult, lends support to this "pulling, flattening" hypothesis.[10]

On the other hand, other coordinated events also occur such as the change of lens power and the coordinated increase in eye size (i.e., most importantly an increase in axial length). The crystalline lens, which averages 45 D during infancy will lose about 20 D of power by age 6.[11,12] In order to compensate for this loss of lens power, the axial length will increase by 5 to 6 mm in that same time frame.[3] (As a rule of thumb, 1 mm of change in axial length equals about a 3 D change in refractive power of the eye.)

Now let me present a suggested mechanism that could account for most of the data.[12,13] As the cross-sectional area of the eye expands, there is an increased pull on the lens zonules and a subsequent flattening of the anterior lens surface (a bit more) and the posterior lens surface (a bit less), thus decreasing the overall lens power. Parenthetically, there also may be a related decrease in the refractive index of the lens that contributes to the reduction in lens power.

Because the incidence of myopia starts to accelerate significantly at about the age of 10,[12] one may ask if there is a decoupling of the previously described coordinated drop in lens power and increase in axial length. As discussed in the chapter on refractive errors, an increased amount of near work (i.e., school work) is associated with a higher incidence of myopia. It is also well known that genetic predisposition also influences myopia incidence because more Asian children than Caucasian children are myopic. Thus, one might hypothesize that the long periods of accommodation that accompany school work (ciliary body contraction), may tend to stretch and weaken the linkage between the enlarging scleral shell and the ciliary body. If this were to happen, then the lens would flatten less during eye growth. Another way of looking at this phenomenon would be to theorize that with the linkage weakened, the restraining effect of the lens-zonule combination on eye growth is also weakened, and an increase in axial length in the myopic student would occur. Many studies demonstrating that the myopic eye has a greater axial length than the emmetropic eye tend to support this idea.[14]

B. RETINAL RECEPTORS

The cone photoreceptors of the retina are responsible for sharp vision under daylight conditions. The denser the cones are packed, the more acute the vision.[1,7] To use a photographic analogy, film with the highest resolution has smaller photosensitive grains, packed tightly, whereas a film with large grains of silver halide yields a coarser picture.

The most sensitive part of the retina is the fovea.* Here the cones are even finer and packed together even tighter. The fovea of the infant eye is packed less than one fourth the density of that of the adult. Furthermore, the synapse density in the neural portion of the retina, as well as in the visual brain, is low at birth. The combination of these two anatomic configurations means that fewer fine details of the retinal image are recorded and sent to the brain.

C. NEURAL PROCESSING

Finally, the nerves that transmit the visual information to the brain, as well as the nerve fibers at the various levels within the brain, are poorly myelinated in the infant. Myelin is the insulating wrap around each nerve fiber. A normally myelinated nerve can transmit nerve impulses swiftly and without static or "cross talk" from adjacent nerves. To use a computer analogy, one might think of the infant brain as being connected with poorly insulated wires. Therefore, sparks, short circuits, and static all slow down or interfere with perfect transmission, and only the strongest messages get through. Figure 1.4 represents an appropriate analogy. The face of Albert Einstein is shown on a computer screen with larger and larger pixels. The infant's early vision might be akin to the picture with the biggest block pixels. With the maturation of the brain processing elements, the neurologic

---

* F.W. Campbell quotes Stuart Ansti's clear analogy of how the fovea functions: "A retina with a fovea surrounded by a lower acuity periphery can be compared to a low magnification finder telescope with a large field of view which will find any interesting target and then steer on to it a high powered main telescope, with a very small field which could examine the target in detail." Campbell FW (1968). The human eye as an optical filter. *Proc. IEEE* 56(6): 1009–1014

FIGURE 1.4   A computer display of the face of Albert Einstein, with pixels of different size. (From V. Lakshminarayanan et al: Human face recognition using wavelengths. In *Vision Science and Its Application,* Santa Fe, February 4, 1995, Optical Society of America p. 18, with permission.)

equivalent of pixel size gets smaller and more details can be registered. Thus, the photographic film grain size in the retina and the equivalent of pixel size in the brain processor both get smaller as the child grows. My colleague, V. Lakshminarayanan, who created Figure 1.4, speculated that the immaturity of the infant's memory capacity may be one reason why its visual images have less detail. In other words, the coarseness of the infant visual system does not overtax the immature memory system.

2. RELEVANT EARLY PHYSIOLOGY

Experiments with infants teach us that good color vision does not appear for about 3 months. The infant will also take longer to "make sense" out of its retinal image. The infant must stare for relatively long periods (1–3 minutes), blinking very rarely during this period. This brings up a fascinating difference between the tearing system of the infant and the adult.[1,7]

If an adult were to blink as infrequently as a baby, the adult's cornea would tend to dry and small defects would develop that act as fine cracks on the optical surface of the cornea. Such dry cracks tend to degrade the retinal image. This does not happen to the infant cornea. It is believed that the tears of the infant have a special composition (as yet unknown) that both resists evaporation and seems to adhere more firmly to the surface of the cornea. If the infant tear formula is ever discovered, it would make an excellent artificial tear preparation for older patients suffering from dry eyes.

3. RECOGNIZING FACES

The remarkable thing about the infant's eyesight is that the relatively poor level of resolution just described still allows the infant to recognize different faces and different facial expressions. We know this to be true in some newborn infants, who can accurately imitate the expressions of an adult, as seen in Figure 1.5. You might think of the baby using its own face as a canvas to reproduce the facial

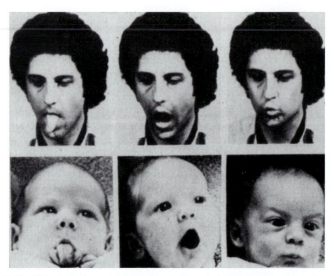

FIGURE 1.5   This photo shows a recently born infant mimicking the expression of psychologist Dr. A. Meltzoff. The baby is obviously able to perceive the different expressions in order to mimic them. (From Klaus MH, Klaus PH: *The Amazing Newborn.* Addison Wesley, Reading, MA, 1985, p. 86, with permission.)

expression of the onlooker. Can you think of a better way to establish a human bond?

Experiments with infants demonstrate that they prefer looking at faces or pictures of faces to any other objects. By about 6 weeks, they can zero in on specific features of the face. For example, they can lock in on their mother's gaze. By age 6 months they can also recognize the same face in different poses. As a matter of fact, they are experts at recognizing a face, be it upside down or right side up until the age of 6 years. Oddly enough, after age 6, they actually lose their skill at quickly recognizing upside down faces.[7]

At age 6 months, a very unusual change has started to take place in the optics of the eye. Gwazda et al.[15] found that a significant amount of astigmatism develops in 56% of the infants studied. This condition will remain for only 1 to 2 years.[15–18] Figure 1.6 gives the reader an idea of how this amount of astigmatism can distort the image of a round ball. The round ball in the picture seems pulled into the shape of a football. To take it a step farther, astigmatism tends to elongate features and exaggerate lines. Astigmatism can be described as to the amount of distorting elongation in diopters and the direction of the distortion is described in degrees from the horizontal (0° to 180°) axis. Figures 1.7A and B (see color insert) are a hypothetical construction of how the baby's astigmatism might change the retinal image of its mother's face. Note the manner in which the mouth and eyes are drawn out to be more linear (much as a line drawing). Note too, that the elongation in this picture is in an oblique orientation, rather than horizontal or

FIGURE 1.6   An illustration of how vertical astigmatism can distort the image of a round ball.

vertical. This oblique direction is an example of an astigmatism with an axis of 45° (45° from the horizontal direction).

Let's speculate again. The suggested line drawing rendition of the mother's face (which may also represent the first stage of visual processing in the adult) reminds me of the faces created by the famous pantomime Marcel Marceau. In Figure 1.8 (see color insert) Mr. Marceau is shown using white make-up to obscure all the creases in his face. He then uses dark lines to only outline his mouth and eyes. Figures 1.9 and illustrates some of the expressions that he presents. You might think of the mime as creating different line drawings of the face. What is astonishing is that although made of only a few dark lines, the mime can recreate most human expressions. It seems reasonable to imagine that the mime presents faces similar to that seen by the young child or found in a child's drawing. Look at the faces in children's drawings in Figures 1.12, and 1.13. The faces have no texture, no shadowing, no creases, only a line mouth, circles for eyes, and occasionally a dot nose. Is it not then possible that infant astigmatism helps represent faces as line drawings to the infant visual system? Line drawings also save

FIGURE 1.9   Some different expressions of Marcel Marceau. Because of his use of make-up, his face resembles a line drawing. Yet the various expressions are easily discerned. (From Martin B: *Marcel Marceau, Master of Mime.* Paddington Press Ltd., London, 1978, with permission.)

memory storage space, which would be an advantage for the small infant brain.* The illustration in Figure 1.10 makes this point in a different way. The face of Albert Einstein is shown with a complete gray scale on the left and is shown as a line drawing (only black and white) on the right. The line drawing requires much less computing power than a face with texture, and would be more compatible with the child's immature processing system.

FIGURE 1.10   A series of computer simulations of the face of Albert Einstein with an extensive gray scale on the left (16 gray scale) and only a 2 gray scale (i.e., similar to a line drawing) on the right. (From Lakshminarayanan V., et al: Human face recognition using wavelengths. In *Vision Science and Its Applications,* February 4, 1995, Santa Fe, Optical Society of America, p. 167, with permission.)

* This idea was suggested by David Marr in a seminal article, "Early processing of visual information." (*Phil. Trans. R. Soc. London, Seri. B* 275:483–524, 1976.)

4. LINE ORIENTATION RECEPTORS

As noted earlier, the amount of astigmatism can rise to a level of over 2 diopters in the first year of life in many infants. The orientation of the distortion is usually horizontal (180°) initially. In the course of the next 2 years, the axis of distortion rotates to the vertical, and the amount of the astigmatism diminishes. This slow rotation of the axis of exaggeration could help turn on different groups of brain cells that will become sensitive to features in the retinal image with different tilts. In fact, the discovery of these brain cells with orientation selectivity led to a ground-breaking understanding of the functional architecture in the higher brain. Torsten Weisel and David Hubel, working in their laboratory at Harvard Medical School late one night in 1958, had implanted electrodes in the visual cortex of an anesthetized cat in order to record cortical cell responses to patterns of light, which they projected onto a screen in front of the cat. After 4 hours of frustrating work, the two scientists put the dark spot slide into the projector, where it got jammed. As the edge of the glass slide cast an angled shadow on the retina, the implanted cell in the visual cortex fired a burst of action potentials. Torsten Weisel described that moment as the "door to all secrets." The pair went on to prove that cells in the cortex responded only to stimuli of a particular orientation. Similar responding cells were all located in the same part of the cortex. This work opened up the area of how and where the brain encodes specific features of the retinal image. Fittingly, Drs. Hubel and Weisel were awarded the Nobel prize for medicine in 1981.[19]

5. MONITORING OTHER'S EYE MOVEMENTS

The British psychologist Simon Baron-Cohen, in his book *Mindblindness** suggests that a major evolutionary advance has been the human's ability to understand and then interact with others in our social group (i.e., play social chess). He further suggests that we accomplish this social intelligence primarily by following the eye movements of others, and we start at a young age. For example, an infant of 2 months starts concentrating on the eyes of the surrounding adults. The infant has been shown to spend as much time on the eyes as all the other features of an observer's face.

By 6 months, the infant will look at an adult's face two to three times longer if the face is looking at the infant than if it is looking away. We also know that when the infant achieves eye contact, a positive emotion is achieved (i.e., the infant smiles). By age 14 months, infants start to read the direction that an adult is looking. It will turn in that direction, and then continue to look back at the adult to

---

* In his book *Mindblindness: An Essay on Autism and Theory of Mind* (MIT Press, Cambridge, MA, 1995) the author ferrets out the key features of "eye following" in the normal child by comparing it to the autistic child.

check that both are looking at the same thing. By age 2 years, normal infants can read fear and joy from eye direction and facial expression.

6. RECOGNIZING MOVEMENT

An infant is capable of putting up its arm to block a threatening movement. This act tells us that the infant both appreciates movement and the implied threat of this particular movement.[5,20] Admittedly, the infant cannot respond if the threat moves too quickly, probably because the immature myelinization of its nerves slows down all the neural circuits. Nevertheless, a definite appreciation of movement and threat exists. To register movement accurately, the infant retina probably records some object at point A. That image is then physiologically erased (in the brain and/or retina) and the object is now seen at point B. This physiologic erasure is important or else movement would produce a smeared retinal image. Researchers think that the baby probably sees movement as a smoothed series of sharp images, and not smears.[21] This hypothesis gains support from other experiments that demonstrate that at a very early age, the baby can appreciate the on/off quality of a rapidly flickering light. It seems logical that the movement of an image across the retina (with the inherent erasures) is physiologically related to the rapid on/off registration of a flickering light.

A movie hero like special agent James Bond is always aware of the presence of danger out of the "corner of his eye." Technically, this maneuver is known as the *foveal reflex*. That is, a new target just seen out of the corner of your eye (peripheral retina) stimulates the eye muscles to rotate the eye so that the sensitive fovea is aimed at the new object. By 2 weeks of age, the infant demonstrates this very same reflex.

7. RECOGNIZING THREE DIMENSIONS

The young child also has other sophisticated visual skills. One Sunday morning many years ago when my middle son was only 2 years old, I took him with me on hospital rounds. We stopped at my office, where I changed into my white hospital coat. Just as we were leaving, a large cockroach scurried along the floor. When the boy saw it, he started screaming. In an attempt to minimize his fear of a live bug, I quickly pulled out the stereopsis test that we show to our patients. It is a flat picture of a fly. If you look at it through special Polaroid glasses, the fly takes on a three-dimensional quality. When he looked at the picture without the special glasses, he smiled. However, when he looked at the fly again with the glasses, he started screaming. The realism of the three-dimensional fly reminded him of the cockroach. Obviously, this stereoscopic capability is present at a very young age and helps the infant recognize certain real-life dangers.

8. BLOCKAGE OF VISUAL DEVELOPMENT
(AMBLYOPIA)

As has already been observed, the child's visual apparatus is in constant growth as it is accumulating new visual information, somewhat like the way a small public library in a young town grows as the town enlarges. In fact, let's make the analogy more realistic and give the young town two librarians (i.e., eyes) working in two different locations, ordering books for the main library (i.e., visual cortex). The town fathers also have a library contingency plan. If one of the librarians is more able than the other, the able librarian holds the dominant position. In other words, the main library is primarily run by that librarian. In essence, the second librarian is not only a helper, but a backup, or replacement librarian. In the event that the dominant librarian gets sick or actually leaves, then the second librarian takes over. The visual system seems to work with a similar strategy. If one eye turns in or out (i.e., a squint), and so is not pointed in the same direction as the dominant eye, it will not develop the same degree of fine visual acuity. This condition of poorly developed central vision is known as *deprivation amblyopia.* If the diagnosis is made before the age of 10,[22–24] the dominant eye is covered with a patch for a few months. This treatment seems to allow the brain circuitry connected to the amblyopic eye to develop further and achieve a visual acuity almost equal to the dominant eye. After the patching treatment is concluded, although the turned eye remains turned, one of three visual results will emerge. The vision in the turned eye will slip back to the pretreatment level and the dominant eye will take over as before. However, the treatment has forced new circuits in the brain to be created. Therefore, if the dominant eye is ever damaged or lost, the vision in the amblyopic eye can be improved to the level attained from the patching treatment. A second possible result of the patching treatment is to have a condition known as an alternating squint. In such a case, each eye retains good visual acuity. However, only one collects detailed information at any one time, that is, the dominance seems to alternate from visual task to visual task. Such a condition may evolve whether the eye continues to be crossed or is straightened surgically. The final possibility is that after patching and a successful eye straightening surgery, both eyes retain good visual acuity and work together simultaneously. In sum, with deprivation amblyopia, there is a window of time for the development of good visual acuity in a turned eye (or in an eye with uncorrected severe farsightedness or astigmatism) if appropriate treatment is applied.

Interestingly, the window of time for good visual acuity narrows down to a few months of life if one eye or both eyes are blind due to a congenital cataract or a congenital corneal opacity. In either case, an extremely blurred image is formed on the retina. Under these circumstances, the body appears to use another strategy. If vision is not restored surgically or by some rare natural occurrence well before the end of the first year of life, then no corrective surgical procedure, later in life, will end in good visual acuity. To clarify, if a cataract develops later in childhood, after that eye has developed good visual acuity, then a surgical correction can restore good vision. On the other hand, visual acuity in an eye with a dense congenital cataract (if the other eye is normal) will never reach a high level if the surgical removal of the cataract is performed after the first few months of life.[25]

Professor Richard L. Gregory, in his book, *Odd Perceptions,*[26] tells the sad story of a man who was blind since birth from congenitally opaque corneas in both eyes. At age 52 he underwent a successful corneal transplant in one eye, which allowed a sharply focused image to be placed on the retina. Unfortunately, after the surgery, he could only see things that he already knew through touch, and therefore, remained blind to things that he had never previously explored. When Professor Gregory heard that the man had always been interested in machine lathes, he took the man to an industrial museum. The man ran his hands over the lathe with his eyes shut. Then he stood back a little and opened his eyes and said, "Now I've felt it, I can see it."

This case typifies the events that occur when an adult patient, who has been blind since birth, undergoes eye surgery later in life. The anticipation of a life of normal vision is never fulfilled. Unhappily, case reports show that depression and occasionally suicide often follow the "successful operation." Of course, not all cases are simple to evaluate. I can clearly recall being convinced by an adult patient with poor vision from cataracts since early childhood to perform a cataract extraction. I had been aware of Professor Gregory's patient, but I thought that the dim vision that my patient had was enough to allow him to develop those circuits in the brain as a child. Following the technically successful operation, his vision improved only a bit. Unfortunately for him, his visual acuity improved just enough so that he lost his government benefits as well as membership in the Massachusetts Association of the Blind. A very depressed patient ultimately returned to me and begged me to recertify him as blind. Although he could identify a few large letters on the vision chart, he truly functioned as a blind man. Imagine the surprised and mixed reaction in the waiting room of our clinic when I shook his hand and said, "Congratulations, I will certify that you are legally blind."

The point of this discussion is that there is a critical period early in visual development in which innate neural wiring and visual experience must interact to permanently imprint visual skills. Broadly speaking, heredity and environment (nurture and nature) work together to provide the hardwiring and the software (visual experience). However, the time for this process to work to full potential is limited.

9. WHAT CHILD ART TEACHES US ABOUT VISUAL DEVELOPMENT

How, aside from clinical vision testing, can we follow the visual development of the child? We can look at his or her drawings. By age 3, when hand coordination can control a crayon or a stick, children all over the world make drawings. Rhoda Kellog,[27] an investigator who has analyzed over 100,000 children's drawings from many cultures, tells of a Nepalese 3-year-old who, when given a crayon and a piece of paper for the first time, immediately created drawings identical to those of his counterparts in the United States. What also appears universal is the joy and contentment associated with drawing. Every mother knows that the bored child is usually converted to the happy child when given crayons or paint and a piece of paper. Clearly, the ability to draw and the joy involved in drawing are innate.

A. DRAWING FACES

The young child starts off by experimenting with circles. Very quickly these circles become faces. In a short while, the faces take on radiating arms and legs. This pattern is called a *mandela*. It is drawn by most children. There is an interesting illusion, called the Ehrenstein figure, which consists of radiating lines against a light background. The illusion produces the perception of a disk that seems to connect the lines. (Figure 1.11). The almost universal drawing of the mandela by children and this illusion suggest that the circular shape may be innate to our visual circuitry. Later the child will start to concentrate on drawing a face with a body. Figure 1.12 is an example of the first human renderings of 3- and 4-year-olds. Note that the extremities start as radiating lines. In time, the lines representing the legs are placed at the bottom of the face and the arms at the side of the face. Of course, not all children progress at the same rate. Figure 1.13 rep-

FIGURE 1.12   The progression of the drawings of the 3- and 4-year-old so that the arms radiate from the side of the face and the legs from the bottom. (From Kellog R: *Analyzing Children's Art.* National Press Books, Palo Alto, CA, 1970, with permission.)

FIGURE 1.13   The picture of a man drawn by two different 5-year-olds. Note that one child has advanced to the point of drawing a torso. (From Brittain WL, Lowenfeld V: *Creative and Mental Growth*. Macmillan, New York, 1982, p. 57, with permission.)

resents a man drawn by two different 5-year-olds. The more advanced of the 5-year-olds has incorporated the concept of the human torso into the drawing. Using the torso, the child will then branch to other living forms. In Figure 1.14, a human form has been modified to become a cat, and in Figure 1.15, the human form with torso is slightly altered to depict a tree.

FIGURE 1.14   A first grader has modified a human form to draw a cat. (From Snow AC: *Growing with Children Through Art*. Rheinhold Book Corp. New York, 1968, with permission.)

FIGURE 1.15   Children 5 to 7 years of age modify the human form to depict trees. (From Kellog R: *Analyzing Children's Art.* National Press Books, Palo Alto, CA, 1970, p. 29, with permission.)

What can we learn about the developing visual system through such drawings? The first message is that human faces and human forms are the first objects we draw. This connectedness with human faces, which was first seen in the new-born's ability to imitate facial expressions, continues to take top priority (Nature). The next message is that adults teach us how to see other objects (Nurture). They teach us to give meaning and relevance to the retinal image. Children's drawings can demonstrate this point. In Figure 1.16, we see examples of the free-spirited, almost abstract drawings of people, followed by the teacher-influenced drawings of the same people by children of the same age.

B. THE INFLUENCE OF CULTURE ON CHILD ART

The teacher's influence forces the drawing to look more like a socially acceptable depiction of people. Figure 1.17 represents a similar message using animals. Again, the teacher-influenced drawings are less abstract and more accurate. Finally, in Figure 1.18, there is a similar comparison of houses. The child must learn the socially agreed upon renditions of people, animals, and objects. Perhaps a story will better explain how powerful social influence is on

Humans that most teachers would like (five to seven years)

Humans that few teachers would like (five to seven years)

FIGURE 1.16 The influence of society as represented by the teacher in the drawings of people done by 5- to 7-year-olds. The drawings below were made prior to teacher instruction. (From Kellog R: *Analyzing Children's Art.* National Press Books, Palo Alto, CA, 1970, with permission.)

vision. When the famous anthropologist, Malinowski, visited the Trobriant Islands, he was surprised to learn that it was taboo to say that two siblings looked alike or that they resembled their mother, even though the resemblances were very obvious to Malinowski. In making comparisons, one was only allowed to note that a child looked like its father. After openly discussing the issue with many natives, Malinowski truly felt that the people were not lying when they reported no facial resemblance between two brothers. They had simply been taught how to see, and had developed their version of reality. Thus, the workings of the visual pathways had been biased by the local psychosocial influences. A. I. Hallowell[28] put this idea succinctly when he wrote, "The world indeed looks the way a people have learned to talk about it." Thus, along with all the innate visual functions that have been discussed, are the equally important cultural interpretations of what we see. These are formed in the first 10

FIGURE 1.7 The normal young mother's face on the right has been distorted horizontally and blurred on the left by an astigmatic lens. It is suggested that such a distortion, as seen by a baby, mimics a line drawing rendition of the mother's face.

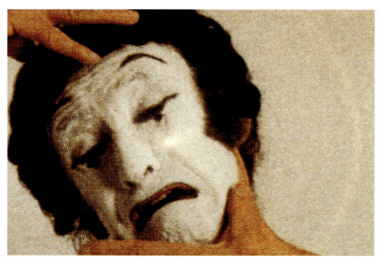

FIGURE 1.8 A photograph showing the famous pantomimist Marcel Marceau making up before a performance. Note the outlining of the mouth and eyes against a matte white face. (From Martin B: *Marcel Marceau, Master of Mime.* Paddington Press Ltd., London, 1978, with permission.)

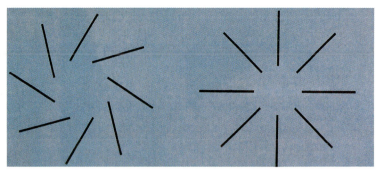

FIGURE 1.11 The Ehrenstein figure *(left)* gives the impression of a disk connected to radial dark lines, much as a mandela. No such disk is seen against the dark nonradial lines in the figure on the right.

Animals that most teachers would like (five to seven years)

Animals that few teachers would like (five to seven years)

FIGURE 1.17 The influence of society as represented by the teacher in the drawings of animals done by 5- to 7-year-olds. As in Figure 15, the drawings below were made prior to teacher instruction. (From Kellog R: *Analyzing Children's Art.* National Press Books, Palo Alto, CA, 1970, with permission.)

years of life. Perhaps Anais Nin said it even better when she noted, "We don't see things as they are, but as we are."[29]

SUMMARY

A. SOCIAL SEEING

Early in the chapter it was shown that the visual system of the infant is immature. Yet, some infants can recognize and respond to adult facial expressions on the first day of life and follow their mothers' glances by 6 weeks of age. Clearly, infants' top priority is to maintain a social relationship with their mother or other caring adults. This idea was put succinctly by the language expert, Pinker,[30] "Most normally developing babies like to schmooze." As the child grows, he or she learns to see people and objects in the way society and culture demands and communicate in the expected manner, that is, "We don't see things as they are but as we are." Is it possible that the immature eye and visual system actually facilitate socially biased seeing? Perhaps the smaller, simpler retinal images, along with the less sophisticated brain processing, pre-

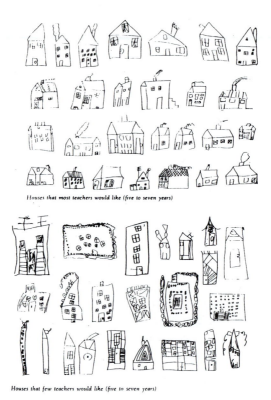

Houses that most teachers would like (five to seven years)

Houses that few teachers would like (five to seven years)

FIGURE 1.18 The influence of society as represented by the teacher in the drawings of houses done by 5- to 7-year-olds. As in Figures 15 and 16, the drawings below were made prior to teacher instruction. (From Kellog R: *Analyzing Children's Art.* National Press Books, Palo Alto, CA, 1970, with permission.)

vent the many other details of life from confusing the key message, that is, social interaction takes top priority.

B. AMBLYOPIA

It was noted that visual development can be blunted in an eye that is turned, has a severe refractive error, or a congenital opacity. In such cases, visual acuity (in the weaker eye) can be brought to a higher level if the good eye is patched for a few months and the refractive error in the weaker eye corrected early. One way to picture the mechanism of this reversible decrease in vision is to assume that the visual system has some inherent interference static built into the circuitry of the amblyopic eye. Perhaps an example will clarify the concept of static or electronic jamming. While I was working as an army doctor, a young soldier came to my office complaining of insomnia. I assumed it was due to the severe itching from fungal infection of the feet and prescribed a special ointment. When the soldier returned, he was all smiles. He told me that he was sleeping better. I assumed that he slept better because his feet had stopped itching. Not true. "You see, Doc," he said, "although the lights are shut off at 10 P.M. in our barracks, some of the soldiers continue to play their radios and that keeps me awake. After getting nowhere with politeness, I took matters into my own hands. I secretly hooked up a radio jamming device to the main light switch." When the "on duty officer" shut off the lights, he activated the jammer and static drowned out the music. "Well in no time, the soldiers turn their radios off in disgust and I get a good night's sleep." In a way not fully understood, patching the good eye seems to disconnect the jamming or static fed into the amblyopic eye and vision is allowed to develop.

On the other hand, if one or both eyes cannot form a useful image on the young retina because of cataracts or opaque corneas, only a corrective operation at any early age can ensure normal visual development. If curative surgery is not performed on the young eye then the brain area reserved for visual processing is either closed down or used to expand touch or sound processing. In support of this concept is the fact that people blind since birth use the traditional vision area of the brain to process information from their finger tips.[31]

C. CHILD ART

A key ingredient of art lies in the human ability to substitute an artificial rendition of the real event, in other words, pretending that an artificial version of the event is the real thing, and then manipulating it mentally. The ability to draw may be related to the ability to produce graphic models of events and processes. Thus, the ability to produce such visual displays can be seen as an important component of human technological progress. If in fact drawing ability can be linked to human progress, then it is not surprising that children all over the world have an innate ability to draw.

Children can both create pictures and play "let's pretend" at an early age.* Clearly, children can exert more control over their "play world" than the adult-controlled real world. We know that their interest in mastery over their world starts at an early age. Experiments done on young infants demonstrate that such a mastery gives the infant a sense of satisfaction. Specifically, 3-month-old infants were placed in a crib in which the movement of a mobile, hanging over the crib, was connected to a device that the baby could manipulate. Once the baby discovered the connection, it would smile whenever its controlling manipulation moved the mobile.[32] Interestingly, the infant demonstrated sadness if it failed to move the mobile. This seems to suggest that we might be born with a desire to master our environment. One way to achieve this end is to re-create our environment as in a drawing.

---

* Infants can identify distinctive features in pictures. Yet they obviously do not confuse pictures with the real things, since they never try to eat pictures of cookies.[20]

REFERENCES

1. Boothe RG, Dobson V, Teller DY: Post natal development of vision in human and non human primates. *Ann. Rev. Neurosci.* 8:495, 1985.
2. Collins D: *The Human Revolution: From Ape to Artist.* Phaidon Publishing Company, London, 1976.
3. Larsen JS: The sagittal growth of the eye. Ultrasonic measurements of axial length of the eye from birth to puberty. *Acta Ophthalmol.* 49:872, 1971.
4. Teller. DY: First glances: The vision of infants. *Invest. Ophthalmol. Vis. Sci.* 38(11): 2183–2203, 1997.
5. Fantz RL: Visual perception from birth as shown by pattern selectivity. *Ann. N.Y. Acad. Sci.* 118:793–814, 1965.
6. Green DG, Powers MK, Banks MS: Depth of focus, eye size, visual acuity. *Vis. Res.* 20:827, 1980.
7. Reynolds CR, Fletcher F, Janzen E: *Handbook of Clinical Child Neurophysiology.* Plenum Press, New York, 1989.
8. Inagaki Y: The rapid change of corneal curvature in the neonatal period and infancy. *Arch. Ophthalmol.* 104:1026–1027, 1986.
9. Insler MS, Cooper HD, May SE, Donzis PB: Analysis of corneal thickness and corneal curvature in infants. *CLAO J.* 13:182–184, 1987.
10. Tucker SM, Enzenauer RW, Levin AV, et al: Corneal diameter, axial length and intraocular pressure in premature infants. *Ophthalmology* 99:1296, 1992.
11. Wood ICJ, Mutti DO, Zandnik K: Crystalline lens parameters in infancy. *Ophthalmol. Physiol. Opt.* 16:310–317, 1996.
12. Mutti DO, Zadnik K, Fusaro RE, et al: Optical and structural development of the crystalline lens in childhood. *IOVS* 39:120–134, 1997.
13. Hofstetter HW: Emmetropization-biological process or mathematical artifact? *Am. J. Optom. Arch. Am. Acad. Optom.* 46:447–450, 1969.
14. Goss, DA: Development of the ammetropias. In *Borish's Clinical Refraction* (Benjamin, WJ, ed.). W.B. Saunders, Philadelphia, 1998, pp.52–56.
15. Gwazda J, Scheiman M, Mohindra I, Held R: Astigmatism in children: changes in axis and amount from birth to six years. *Invest. Ophthalmol. Vis. Sci.* 25:88, 1984.
16. Atkinson J, Braddick O, French J: Infant astigmatism: It's disappearance with age. *Vis. Res.* 20:801, 1980.
17. Banks M: Infant refraction and accommodation. *Int. Ophthalmol. Clinics* 20:205, 1980.
18. Mohindra I, Held R, Gwazda J: Astigmatism in infants. *Science* 202:329, 1978.
19. Strickland C: Torsten Weisel, Winner of 1981 Nobel Prize for Vision Research. *Argus* (publication of the American Academy of Ophthalmology, San Francisco) Jan. 1995, pp. 8,9.
20. Bower TG: *The Perceptual World of the Child.* Harvard University Press, Cambridge, MA, 1997.
21. Tronick E: Simultaneous control and growth of the infant's effective visual field. *Percept. Psychophys.* 11:373–376, 1972.
22. Vaegan-Taylor D: Critical period for deprivation amblyopia in children. *Trans. Ophthalmol. Soc. U.K.* 99:432, 1979.
23. Von Noorden GK: New clinical aspects of stimulus deprivation amblyopia. *Am. J. Ophthalmol.* 92:416–421, 1981.
24. Daw NW: Mechanisms of plasticity in the visual cortex: The Friedenwald Lecture. *Invest. Ophthalmol. Vis. Sci.* 35:4173, 1994.
25. Birch EE, Stager DR: The critical period for surgical treatment of dense unilateral cataract. *Invest. Ophthalmol. Vis. Sci.* 37 (no.3):1532, 1996.
26. Gregory RL: *Odd Perceptions.* Routledge, New York, 1986, p. 176.
27. Kellog R: *Analyzing Children's Art.* National Press Books, Palo Alto, CA, 1970.
28. Hallowell AI: *Culture and Experience.* University of Pennsylvania Press, Philadelphia, 1955, pp. 30–32.

29. *Physics Today,* Jan. 1995, p. 55.
30. Pinker S: *The Language Instinct.* Penguin Books, New York, 1994, p. 267.
31. *Discovery Magazine,* Vol 17, No. 7, July 1996 (Work of N. Sadato described).
32. Watson JS: Smiling, cooing and the game. *Merrill-Palmer* 18:323–329, 1972.

2

THE IMAGE OF THE
ADULT HUMAN EYE

1. Tuned to Visible Light Waves
   A. *Role of the Cornea*
   B. *Role of the Crystalline Lens*
   C. *Accommodation*
   D. *Role of the Retina*
2. Optical Aberrations
   A. *Light Scattering*
   B. *Natural Defenses Against Light Scattering*
   C. *Chromatic Aberrations*
   D. *Spherical Aberration*
3. Field of Vision
Summary
   A. *A Compromise of Eye Function*
   B. *The Aging Eye*
   C. *Evolution of Ocular Components*

The image quality of the human adult eye is far superior to that of the human infant, although probably inferior to certain predator birds. Its wide focusing range is smaller than that of certain diving birds, and its fine sensitivity to low light levels is weaker than spiders or animals with a tapetum lucidum. Its ability to repair itself is probably not as efficient as some animals (e.g., newts, which can form a new lens if the original is damaged). Finally, the human eye has the ability to transmit emotional information (e.g., excitement via pupil dilation, sadness via weeping) but with less forcefulness than some fish that uncover a pigmented bar next to the eye when they are about to attack, or the horned lizard that squirts a jet of blood from its eye when it is threatened. Thus, in reading this chapter on human ocular optics, one must appreciate its levels of performance in light of its large spectrum of functions.

1. TUNED TO VISIBLE LIGHT WAVES

When our retinas receive an image of a spotted puppy in a room, what is really going on in terms of information transfer? Light waves from the ceiling light are cast onto the puppy. Its body reflects and scatters the light waves onto our eyes. In a sense, information about the puppy has been encoded into visible light waves. The optical elements of the eye focus the encoded light waves onto our retina, as a map of bright and dim colored dots, known as the *retinal image.* Nerve signals report the retinal image to the brain. In the brain, the nerve signals are re-created into the impression that a real puppy is in the room.

One might liken the function of the eye to that of a radio, which receives radio waves carrying Beethoven's second symphony. The specific station broadcasting the symphony beams it out on a specific radio wavelength. Then your radio receives the radio waves and the speaker reconverts the musical sounds. If the eye is to receive and process visible light, it must be constructed so as to be able to be tuned to the wavelengths of visible light. The physicist would substitute the term resonate for tune. Literally, resonate means to re-sound another time. Let us look at a simple example of resonance. An opera singer can make a wine glass hum when the frequency (or wavelength) of the note is the same as the natural frequency of the glass. What is the natural frequency of anything? The answer has to do with composition and size. Organ pipes of different lengths each resonate at different frequencies and wavelengths. By changing the length of the antenna on your car radio, you can pick up the frequencies of different stations. The best receiver for a specific wavelength (frequency is the reciprocal of wavelengths) is either physically the same size as the wavelength, or a precise number of wavelengths, or a precise fraction ($1/4$, $1/2$) of the size of the wavelength. Therefore, optical theory demands that the size of the key components of the eye should be the size of a wavelength of visible light or some number *(n)* times the size of, or a fraction of the size of the wavelengths of visible light (as well as made of a resonating material). Let us see if that is true.

A. ROLE OF THE CORNEA

The human cornea is a unique tissue. First, it is the most powerful focusing element of the eye, roughly twice as powerful as the lens within the eye. It is mechanically strong and transparent. Its strength comes from its collagen fiber layers. Some 200 fiber layers criss-cross the cornea in different directions. These fibers are set in a thick, watery jelly called a *glycosaminoglycan.* The jelly gives the cornea pliability. For a long time, no one could come up with a convincing optical explanation for the transparency of the cornea. No one could understand how nature combined tough, transparent collagen fibers (with their unique index of refraction) with the transparent glycosaminoglycan matrix, which had a different index of refraction, and still maintain clarity. Perhaps an everyday example of this phenomenon will help. When you fill a glass from the hot water tap, the solution looks cloudy. If you look closely, you see many fine clear expanded air bubbles (which have a unique index of refraction) within the clear water, which has different refractive properties. On the other hand, cold water appears clear because its air bubbles are very tiny. The normal corneal structure might be considered optically similar to the structure of the cold water (i.e., tiny components with different indices of refraction).

Although the transparency is obvious to all, the physical explanation was a long time in coming. In the 1960s, the prevailing theory suggested that if the collagen fibers were arranged in a perfect crystal lattice, transparency would be the result. Then Dr. Jerry Goldman, a young ophthalmologist studying corneal disease at Harvard, examined the histologic structure of the shark cornea. Yes, the shark cornea was quite clear, but its collagen fibers were arranged in a random fashion. The lattice theory predicted that the shark cornea should be as opaque as a piece of white paper. What was happening? Was Dr. Goldman's histologic specimen incorrectly prepared? He needed help, so he innocently called the central operator at MIT and asked to be connected with a light scattering expert. That fortuitous call brought a professor of physics by the name of George Benedek over to Dr. Goldman's laboratory. Professor Benedek was ultimately able to prove that if the spaces composed of glycosaminoglycan as well as the size of the collagen fibers were smaller than one-half wavelength of visible light, the cornea will be clear, even if it were arranged in a random fashion.[1] It should be noted that an orderly arrangement of the fibers also helps maintain transparency.

Another way to put it is to say that the cornea is basically transparent to visible light because its internal structures are tuned to the size of a fraction of the wavelengths of visible light. Figure 2.1 is an electron micrograph showing the fibers of the human cornea. The black dots are cross-sections of collagen fibers imbedded in the glycosaminoglycan matrix. Note that in this specimen, the fibers are spaced closer than half a wavelength of visible light apart, and the fibers in each of the major layers are arranged in an orderly manner.

This arrangement of corneal fibers serves at least four important functions. First, the arrangement offers maximal strength and resistance to injury.* Second, the arrangement produces a transparent, stable, configured optical element. Third, the fiber arrangement and the birefringent nature of the collagen fibers may help to reduce reflected glare. Fourth, the spaces between the major layers act as potential highways for white blood cell migration if an injury or infection takes place.

---

* A fiber-matrix structure like the cornea has a number of mechanical advantages. For example, fibers separated by a softer matrix cannot transfer stress cracks and tears to neighboring fibers. During bending, the separating matrix prevents fibers from abrading each other. On loading, the matrix can transmit forces around torn fibers.

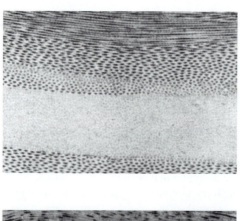

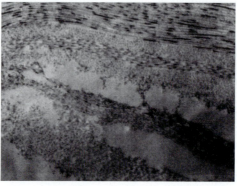

FIGURE 2.1   *Upper photo:* Electron micrograph showing the neat pattern of corneal collagen fibers. The black dots are the fibers cut on end. In this photo the spacing between fibers is less than a wavelength of light apart. *Lower photo:* Large spaces between collagen fibers, as seen in a water-logged, hazy cornea. (Courtesy of T. Kuwabara, M.D., Howe Laboratory, Harvard Medical School. From Miller D, Benedek G: Intraocular Light Scattering. Charles C Thomas, Springfield, IL, 1973, p. 54, with permission.)

B. ROLE OF THE CRYSTALLINE LENS

Have you ever noticed that you see things much better underwater if you wear goggles?* Without goggles, the water just about cancels the focusing power of the cornea,† leaving objects in the underwater world blurred. The goggles ensure

---

* Because the index of refraction of water is greater than in air, objects underwater appear about one-third closer and thus one-third larger[2] than they would in air (i.e., magnification = 1.33×).

† The cornea is a focusing element for two reasons. First, it has a convex surface. Second, it has a refractive index greater than air. Actually, its refractive index is similar to water. Thus, when underwater, the surrounding water on the outside and the aqueous humor inside the eye combine to neutralize the cornea's focusing power.

an envelope of air in front of the cornea, restoring its optical power. If water cancels optical power, how can we explain the focusing ability of the eye lens, which lies inside the eye, surrounded by a fluid known as aqueous humor? The answer is that the focusing power resides in the unusually high protein content of the lens. The protein concentration may reach 50% or more in certain parts of the lens.* Such a high concentration increases the refractive index above that of water and allows the focusing of light. Now we are ready to appreciate the real secret of the eye lens.[1,3]

Normally, a 50% protein solution will be cloudy, with precipitates floating about like curdled lumps of milk in a cup of coffee. However, the protein molecules of the normal lens will not precipitate. In a manner not fully understood, the large protein molecules, known as *crystallins,*** seem to repel each other, or at least prevent aggregation so as to maintain tiny spaces between each other. The protein size and the spaces between them are equivalent to a small fraction of a light wave length. Spaced as they are, you might say that they are tuned to visible light and allow the rays to pass through unimpeded. On the other hand, if some pathologic process occurs, then the protein molecules will clump together and the lens will lose its clarity. When this happens, light is scattered or splashed about, as it passes through the lens. The result is a cloudy lens, known as a cataract, which blurs the retinal image.

A number of years ago, my colleagues and I were working on an optical filter that had the effect of neutralizing the scattered light produced by the cataract. Although we never were able to construct a filter that could be worn in a pair of spectacles, we did prove the principle of the filter in the lab. We were able to take a cataract surgically removed from a patient, mount it on a laboratory optical bench, and neutralize its blurring effect with the filter. How did the filter work? It scattered the light in precisely the reverse way that the cataract scattered the light. Consequently, that individual cataract can be thought of as correcting the filter and creating a sharp image. One newspaper article referred to it as "two wrongs make a right" (Figure 2.2). Unhappily, the laboratory experiment never made it into clinical use for many reasons. The filter would only work for a cataract that never changed, since the most minute progressive change annulled the correcting effect. Next, the filter only worked at a very precise location from the cataract. If the filter was placed in a spectacle that shifted by the slightest amount, the effect would be canceled.

Therefore, as of this writing, the only successful long-term treatment for cataracts is modern cataract surgery, along with artificial lens implantation.

---

* The chemical composition of a focusing element such as the eye lens determines the refractive index. Water has a refractive index of 1.33. As the protein concentration in the water of the lens rises, the index of refraction approaches 1.42.

** Crystallins are large protein molecules ranging in size from 45 to 2,000 kDa.

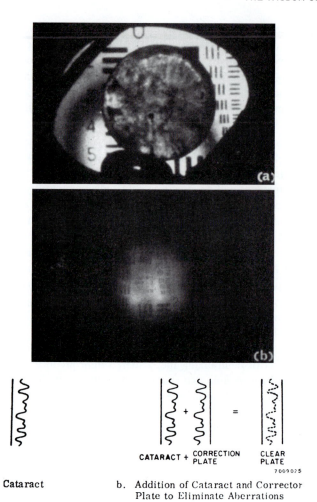

CATARACT + CORRECTION PLATE   CLEAR PLATE

Cataract   b. Addition of Cataract and Corrector
                                Plate to Eliminate Aberrations

FIGURE 2.2   Effect of the complex conjugate of a hologram of a cataract to allow visualization of a target through the cataract. (a) Actual photograph of visual target, covered by cataract and then covered by cataract plus holographic filter. (b) Drawing that illustrates how holographic filter works. (From Miller D, Benedek G: Intraocular Light Scattering. Charles C Thomas, Spring Field, IL, 1973, pp. 115–116, with permission.)

C. ACCOMMODATION

If the emmetropic eye is in sharp focus for this distant world, it must refocus (accommodate) to see closer objects.* For example, the child's range of accommodation is quite large, allowing it to continue to keep objects in sharp focus

* The question "How does accommodation "know" it has achieved the sharpest focus?" seems to be best answered by a sensing system in the brain. However, some have suggested that the system takes advantage of the natural occurring chromatic aberration of the primate eye to fine tune focusing.[4]

from an infinite distance away to objects brought to the tip of the nose. The act of accommodation is quite fast and only takes about one-third second. Our range of accommodation decreases with each passing year so that by age 45, most of us are left with about 20% of the amplitude of accommodation we started with. Why does the eye lens lose refocusing power with age (presbyopia)?* One theory suggests that the lens, like the skin, continually sheds old cells. However, the lens cannot shed these cells into the anterior chamber of the eye (the cells would clog the trabecular filtration meshwork). Thus, the sloughed cells are packed into the center of the lens (nucleus).

With age, the lens enlarges and becomes denser and more rigid. In so doing, it progressively loses the ability to change shape. Parenthetically, it is the cornea of many birds, from pigeons to hawks,† that change shape and is partially responsible for their accommodation. The cornea does not change flexibility with age and so these birds do not become presbyopic. However, there is no "free lunch" in nature. The human eye lens, on the other hand, sits within the eye surrounded by protective fluid, and is far less vulnerable to injury than the cornea.

D. ROLE OF THE RETINA

After light passes through the cornea, the aqueous humor, the lens, and the vitreous humor, it will be focused onto the retinal photoreceptors (rods and cones). Actually, the light must pass through a number of retinal layers of nerve fibers, nerve cells, and blood vessels before striking the receptors. These retinal layers (aside from blood vessels) are quite transparent because of the small size of the elements and the tight packing arrangement. Therefore, we can still say that the size of the important elements of the cornea, eye lens, and retina are of the order of the wavelengths of visible light or a fraction thereof, and can therefore make these organs transparent to visible light.

The bird retina does not have blood vessels. The human has retinal blood vessels that cover some of the retinal receptors and produce fine angioscotomas. The bird retina gets much of its oxygen and nutritive supply from a tangle of blood vessels (the pectin), which is covered with black pigment, and sits in the jelly of the eye, in front of the retina and above the macula (so as to function as a visor). The negative aspect of such a vascular system is vulnerability to a direct blunt or penetrating injury, which could lead to a vitreous hemorrhage and sudden blindness. Obviously this eventuality is less probable in the lifestyle of the bird as compared to that of the mammal or human. Once again, the human eyes can be thought of as having traded safety for improved optical resolution.

---

* More on presbyopia in Chapter 7.

† This was nicely demonstrated on a videotape at an eye research meeting. (Pardue MT, Andison ME, Glasser A, Sivak J: Accommodation in raptors. *Invest. Ophthalmol. Vis. Sci. 37* (no.3) Feb. 16, 1996, p. 725.)

a. Rhodopsin

The rods and cones are made up of a biologic molecule that absorbs visible light and then transduces that event into an electrical nerve signal. The rhodopsin molecule is an example of Einstein's photoelectric effect.* In fact, only one quanta of visible light† is needed to trigger the molecule (i.e., snap the molecule into a new shape).‡ Figure 2.3 (see color insert) is a representation of the molecule rhodopsin. The internal structure of the molecule allows the wavelengths of visible light to resonate within its electron cloud and within 20 million-millionths of a second, induce the change in the molecule that starts the reaction. Let us look at this amazing molecule in more detail.

Probably the earliest chemical relative of rhodopsin is to be found in a primitive purple-colored bacteria, called *Holobacterium halobium*. Koji Nakanishi, a biochemist at Columbia University, in an article titled, "Why 11-cis-Retinal?"[6] (a type of rhodopsin) notes that this bacteria has been on the planet for the last 1.3 billion years.[7,8] Its preference for low oxygen and a very salty environment places its origin at a time on earth when there was little or no oxygen in the atmosphere and a high salt concentration in the sea. Although found in a primeval bacteria, bacteriorhodopsin is a rather complicated molecule, containing 248 amino acids. It is thought that this bacteria probably used rhodopsin for photosynthesis, rather than light sensing. Time-resolved spectroscopic measurements have determined that this molecule changes shape within one-trillionth of a second after light stimulation.[9] This early form of rhodopsin absorbs light most efficiently in the blue-green part of the spectrum, although it does respond to all the colors.[10,11]

To function as the transducer for vision, it must capture light and then signal the organism that the light has been registered. As noted earlier, one molecule needs only one quanta (the smallest possible amount) of light to start the reaction. Even more amazing is the molecule's stability. Although only one quanta of visible light will trigger it, the molecule will not trigger accidentally. In fact, it has been estimated that spontaneous isomerization of retinal (the light-sensing chromophore portion of rhodopsin) occurs once in a thousand years.[5] If this were not so, we would see lights going off every time we had a fever.§ Picture a hair trigger on a pistol that takes only the slightest vibration (but only a very special type of

---

* Some wavelengths of light are powerful enough to knock electrons of certain molecules out of their orbits, and produce an electric current. Albert Einstein was awarded the Nobel Prize for explaining the "photoelectric effect."

† In 1942, Selig Hecht and his co-workers in New York first proved that only one quanta of visible light could trigger rhodopsin to start a cascade of biochemical events eventuating in the sensing of light.

‡ Photoactivation of *one* molecule of rhodopsin starts an impressive example of biologic amplification, in which *hundreds* of molecules of the protein transducer each activate a like number of phosphodiesterase molecules, which in turn, hydrolyze a similar number of CGMP molecules, which then trigger a neural signal to the brain.

§ Such a system has a very high signal-to-noise ratio. Because molecular background noise is not a problem, the system can be seen to be supersensitive.

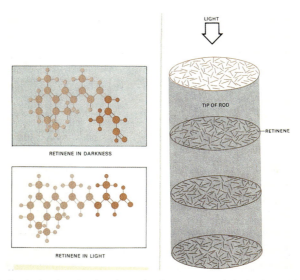

FIGURE 2.3 A drawing showing the molecular structure of rhodopsin and how it changes shape after capturing a photon of visible light. (Modified from Mueller CG, Rudolf M (eds): *Light and Life*. Time Life Books, New York, 1972, with permission.)

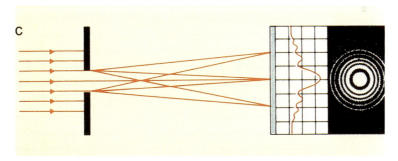

FIGURE 2.5 A diagram demonstrating the appearance of the diffraction disc. (Modified from Jenkins FA, White HE: *Fundamentals of Optics*. McGraw Hill, New York, 1950, with permission.)

FIGURE 2.6   The shape of the eye lens as it sits in the living human eye. Note that both cornea and lens can be photographed because each structure backscatters light to the observer. (The gray curved structure on the right is the cornea. The concave and convex structures on the left represent layers within the eye lens.)

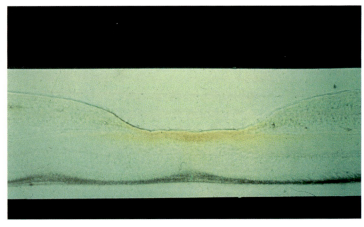

FIGURE 2.7   A cross-section of the center of the retina, illustrating the yellow pigment that absorbs blue light. (From Miller D (ed): *Clinical Light Damage to the Eye.* Springer-Verlag, New York, 1987, with permission.)

FIGURE 2.9   A manmade rainbow. During a bicycle rodeo in Dedham, Massachusetts, the fire department tried to cool off the participants by turning on the hose. The spray from the hose produced the rainbow. (From *Boston Globe,* July 12, 1994, p. 1; photographer John Tlumack, with permission.)

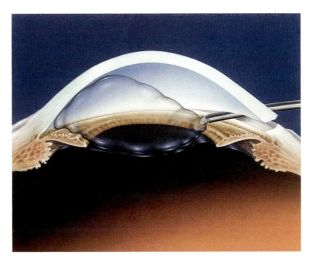

FIGURE 2.11   Three-dimensional drawing of the cornea sitting atop the eye as a small dome, vaulting over the internal structures. This particular picture is from a surgical manual and also illustrates injecting a viscous material into the anterior chamber during cataract surgery (From Pharmacia, Uppsala, Sweden, 1984, with permission.)

vibration) to be activated, and you will have a feeling for the rhodopsin mechanism. As noted, the activating quanta must be of the proper energy level to "kick in" the reaction. That is, the quanta of light must be made of wavelengths of visible light.

b. Receptor Size

The job of a good optical system is to transfer as much information as possible about an object to the corresponding image. Naturally, the retinal image is much smaller than the real objects of the outside world. Since the image is small, the photoreceptors of the retina must be very small to pick up the tiny subtleties within the image (Figure 2.4).

Is there a theoretical limit for the smallness of a receptor? The answer is yes, and the optical limit depends on diffraction. The smallest point of focus of light for any optical system is always surrounded by a small diffraction pattern. Too fine a receptor receiving a diffraction pattern larger than itself would be wasteful. The diameter of the disk of diffraction determines the closest distance that two focused points may be located and not appear to overlap (Figure 2.5 in color insert).[13] In a nutshell, the optimal-sized diffraction disk for a typical human eye, and thus the optimal theoretical diameter of a photoreceptor is between 1 and 2 microns (about 3 times the size of a wavelength of green light). The real diameter of a human foveal cone is close to 1.6 microns. The size of a foveal cone is thus very much tuned to the wavelength of visible light.[12–15]

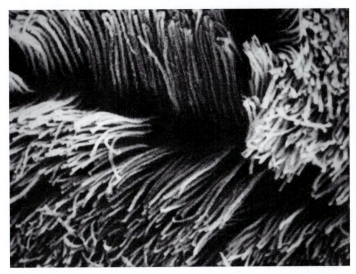

FIGURE 2.4 Electron micrograph showing the cones of the retina to resemble light guides. (From Prause JO, Jensen OA: Scanning electron micrograph of frozen-cracked, dry cracked, and enzyme digested retinal tissue of a monkey and man. *Graefe's Arch Ophthalmol.* 202:261–270, 1980, with permission.)

However, there is a complicating factor that must be kept in mind. The fixating eye, as opposed to a camera on a tripod, is in constant motion. These small movements (called either tremors, drifts, or microsaccades) range in amplitude from seconds to minutes of angular arc. Such movements tend to smear rather than enhance visual resolution. One can only presume that the visual system takes quick, short samples of the retinal image during those smearing movements.[16–18]

Knowing the size of the optimal diffraction pattern of a focused point of light allows us to predict the optimal visual acuity for a human eye.[3] Such an eye should have a visual acuity of 20/20.* In practical terms, a person with 20/20 vision could see a space between two people (i.e., recognizing that there are two people instead of one), which just subtends an angle of 1 minute. To sum up, the unique essence of the vertebrate eye is that the structure of the transparent optical components, the rhodopsin molecule, and the size of the foveal cones are all tuned to interact optimally with wavelengths of visible light.[19,20] It is earth's unique atmosphere and unique relationship to the sun that has allowed primarily visible light, a tiny band from the enormous electromagnetic spectrum of the sun, to rain down upon us at safe energy levels.† Our eyes are a product of an evolutionary process that has tuned to these unique wavelengths and at these levels of intensity.[21,22]

2. OPTICAL ABERRATIONS

Herman von Helmholtz, the famous nineteenth-century German physiologist, in volume I of his *Treatise on Physiologic Optics,* had written that the optical aberrations of the human eye are "of a kind that is not permissible in well constructed instruments." The implication is that the optical design of the human eye would receive low marks if evaluated by someone from the optical industry. If we are simply to compare the optical quality of the living human eye to that of our best cameras and telescopes under ordinary static daylight conditions, then Helmholtz was correct. These optical imperfections will now be discussed.

A. LIGHT SCATTERING

Fingerprints on your spectacle lenses scatter light so that small letters are difficult to read. Raindrops or windshield wiper smears on your car windshield make the reading of street signs difficult. In a similar way, small bubbles from the warm

---

* A few adults in every hundred can see a bit better than 20/20 (i.e., 20/15) and perhaps one adult in a few thousand might see slightly better than that (i.e., 20/10). The incidence of better than 20/20 vision is higher in teenagers.

† The retinal photoreceptors do something very interesting to preempt any light damage that might take place. Every day, they shed their outermost portion of disks of rhodopsin and create an equal number of fresh ones. (Young RW: Biogenesis and renewal of visual cell outer segment membranes. *Exp. Eye Res.* 18:215, 1974.)

water tap give a haze to a glass of water and make it difficult to see the details at the bottom of the glass. These are all examples of light scattering that can obscure the details of any object.

The cornea is, as noted earlier, quite clear, but technically speaking not perfectly transparent. Its composition of fine collagen fibers, loaded into a watery matrix of glycoaminoglycans, and populated by fine cells that swim in the matrix, scatter a small percentage (10%) of incident light and create a very slight haze (Figure 2.6 in color insert). This is unquestionably a flaw, as opposed to the glass optical systems of cameras and telescopes. However, the "imperfect" corneal structure allows for healing. Thus, we can appreciate that a 10% level of light scatter is a fair price to pay for a self-healing system.[1,3]

The lens of the eye is made up of tens of thousands of fine fibers (each a bag of clear protein solution) packed closely together (Figure 2.6). The living lens continually adds outside fibers and packs old cells in its nucleus throughout life, ultimately increasing its volume by a factor of three. The refractive index of the fibers differs from that of the thin spaces between fibers. Thus, the lens scatters about 20% of the incident light. Is there a practical advantage to the fiber structure of the lens? There are many ways that the lens may be injured. These include inflammation from diseases inside the eye, blunt blows from fists or rocks, periods of malnutrition, poisoning, or osmotic upset from systemic disease. In all of these situations, the part of the lens being laid down at the time of the injury or disease loses transparency. This hazy section may also be lumpy in thickness. However, in a short time, new transparent layers will cover the hazy ones, smoothing and reducing the blurring effect of the injury. Again, we now understand that a small level of light scattering due to this fiber layering system is a fair price to pay for a self-repairing lens system.

B. NATURAL DEFENSES AGAINST LIGHT SCATTERING

I may have led the reader astray, suggesting that the eye has no defense against normal light scattering. In fact it does. For example, the birefringent capacity of the collagen fibers in the cornea may cancel out some annoying glare due to the light scattering through a process known as *destructive interference,* somewhat akin to the way polarizing sunglasses cancel annoying glare.

The retina has two defenses against the image-degrading effects of scattered light. To appreciate one of the defenses, you first must know that not all colors (wavelengths) are scattered equally. Blue light is scattered 16 times more than red light by the fine components of eye tissue. Therefore, a defense that can reduce blue scattered light would be disproportionately helpful. Sprinkled throughout the ultrasensitive fovea and its immediate surround is yellow pigment. In Figure 2.7 (see color insert) we see that yellow pigment is very efficient at absorbing the scattered blue light, thus preventing much of it from degrading the retinal image.[3]

The second defense used by the retina involves the positioning of the rods and cones. Each rod or cone functions as a light guide. To enter the guide, light must

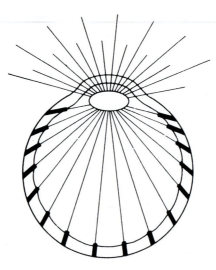

FIGURE 2.8 The direction of the retinal photoreceptors all point to the nodal point of the eye, a point from which the focused rays seem to originate. Scattered light would come from different directions and be less likely to be captured by the photoreceptor light guides. (From Miller D: *Optics and Refraction, A User Friendly Guide.* Gower Medical Publishing, New York, 1991, with permission.)

enter at a specific angle. Interestingly, normally focused light will enter a photoreceptor at a different angle than scattered light. Professor Jay Enoch, former Dean of the University of California School of Optometry at Berkeley, was the first to observe that the photoreceptors of the retina are so directed that they primarily receive focused light, but not scattered light (Figure 2.8).[36]

The dark brown pigment at the rear of the retina and in the choroid absorb any stray light that has passed through the retina and prevent such light from coming back to reverberate among neighboring photoreceptors. None of these defenses is perfect, but all work to reduce the annoyance of scattered light.

The brow and eyelid may also be thought of as blocking annoying glare sources such as the overhead sun. Interestingly, the Asian lid has a double fold and serves as a more effective thicker visor than the Caucasian lid, thereby more effectively blocking the glaring effect of the sun overhead.

C. CHROMATIC ABERRATIONS

The rainbow is produced by millions of tiny round droplets of water vapor that hover over the earth during or after a rain. Each water droplet functions as a tiny, very powerful round lens. Such a powerful lens bends each wavelength of color differently. In Figure 2.9 (see color insert) can be seen a manmade rainbow. During a hot summer day in Dedham, Massachusetts, the fire department turned on a water hose to cool off participants in a bicycle rodeo. The rainbow hovering over the riders was caused by the chromatic aberration of the water droplets. Thus, like Newton's prism, the tiny droplets break up white light into the colors of the rainbow. The phenomenon of strong lenses producing colored fringes around a focused image is known as *chromatic aberration.*[23] The optical components of the eye (cornea and lens), like the fine water droplets, also produce chromatic aberration. This probably sounds odd, because we do not usually see colored fringes around objects. Why? For the most part, the visual processing machinery in the eye and the brain seems to be able to sharpen the edges of our visual impressions and eliminate the fine-colored halos around objects. This manipulation of an image is much like the graphics computer programs that straighten wiggly drawn lines. Although we do not notice the chromatic aberration of the eye, the retina probably senses it and may use the fine-colored fringes around a focused image to help accommodation reach a precise end point.[24,25]

D. SPHERICAL ABERRATION

A major distortion produced by many high-powered optical systems such as the cornea or lens is known as *spherical aberration.* Figure 2.10 shows the results of this aberration. The rays at the edge of the lens are bent more than those going through the center of the lens, creating a smeared focus. The cornea (a strong optical element) is subject to spherical aberration. In Figure 2.11 (see color insert), you can see that the cornea sits at the front of the eye like a small, strongly curved dome. The steeper the dome (shorter its radius of curvature), the more spherical aberration is created. We have known since the time of the French mathematician Descartes that spherical aberration can be controlled by flattening the curvature of the edge of a lens, thereby weakening the focusing power of the lens periphery. Descartes described such a surface as *aspheric.* Most cameras today use lenses with aspheric surfaces. Why isn't the cornea aspheric? It is somewhat. It does get a bit flatter at its periphery so as to merge more smoothly with the sclera. However, there are some real world considerations that make it advantageous to keep the cornea steeply curved. This is only supposition, but in the event

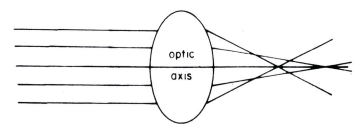

FIGURE 2.10   An illustration of the smearing of sharp focus created by spherical aberration. Note the peripheral rays are bent more than the paracentral rays of light.

of a direct blow to the eye by a stone, the steep protruding cornea can absorb much of the blow, much like a spring. The steeper the cornea, the more it vaults over the rest of the eye, and the greater the spring effect. Such a spring-dampening effect protects the deeper eye structures. How then does the eye satisfy both needs? In daylight with the pupil being constricted, the iris tissue essentially blocks many of the light rays coming through the steep corneal periphery, and effectively cancels most of the spherical aberration. Thus, spherical aberration from the cornea has an important degrading effect only when illumination is dim and the pupil is large. Happily, in dim light, we switch to our rod retinal system in which seeing fine details takes a lower priority than simply seeing large shapes.

The eye lens is also a pretty strong optical element and, therefore, is also vulnerable to spherical aberration. Although an aspheric surface corrects the aberration in cameras and telescopes, nature has chosen a different approach for the eye lens. Recall that refraction or the bending of light can be controlled by either the curvature of the lens surface or by the index of refraction of the composition of the lens. The high index of refraction of the eye lens is the result of a high concentration of protein. The eye lens has a lower refractive index near its edge than at its center. Therefore, the lens periphery has a weaker focusing action and self-corrects spherical aberration, much as an aspheric surface.[26]

Abolition of spherical aberration not only sharpens focus, but can be thought of as concentrating the light at the focus. Concentrating light energy at a focus makes it easier to see a dimly lit object. Therefore, cameras, or creatures with an optical system of minimal spherical aberration, function well in low illumination.

3. FIELD OF VISION

Humans are quite good at noticing events "out of the corner of our eye" (i.e., peripheral vision). In fact, with both eyes open, we can cover about 180° to either side and about 55° up and down. Occasionally, a great sports star demonstrates a wider field of vision.* However, even such a sports star is no match for the mammals, birds, and fish, which have eyes on the side of their heads and have fields of vision 250°.

SUMMARY

A. A COMPROMISE OF EYE FUNCTION

The human eye (which is similar to the monkey eye) is a rather good resolving optical instrument.[12,28–30] As we shall see in the chapter on nonhuman eyes, the

---

* The great American basketball player Bill Bradley[27] was measured to have a 195° horizontal field of vision, that is, he could almost see behind his back.

eyes of certain birds are even better optical instruments, reaching the outer limits of the constraints of the laws of physics. Evolution of better optics has to be balanced against other useful functions such as glare prevention, injury prevention, injury repair, and use of the eyes for nonverbal communication.

B. THE AGING EYE

The components of our eyes usually last a lifetime. This was made clear to me years ago when our eye team did a visual screening of the residents in a housing project for the elderly in greater Boston. The number of people (average age over 80) seeing 20/40 with at least one eye was astonishingly high (over 90%).[31] Of course, I don't want to minimize the disability of those people suffering from cataracts (a prevalence of 9.4 million people in the United States over the age of 65)[32] or age-related macular degeneration (5% prevalence in at least one eye for people over the age of 60.)[32] However, it is important to recognize that a very high percentage of us get through life with good vision. There are number of positive compensations that help the aging eye. Dr. John Weiter,[33] a retinal specialist at Harvard Medical School, points out that the human macula is particularly vulnerable to damage from ultraviolet and blue light. Happily, a yellow material continually builds up in the aging lens that effectively absorbs or scatters away most of these harmful wavelengths, thus diminishing the potential damage to the macula. As noted earlier, the yellow foveal pigment accomplishes a similar function.

One of the more remarkable aspects of the aging eye is the fact that the eye lens continually acquires new layers of fibers, becoming both progressively thicker and steeper. These changes would normally lead to an increase in lens-focusing power and a tendency toward nearsightedness in older eyes. In fact, this does not happen universally because the index of refraction of the outer portion of the lens decreases in a perfectly compensatory fashion,[34] so that the lens power usually stays constant. Frankly, I take this finding to be uplifting. It's nice to think that nature's compensatory processes are not exclusively concerned with those of childbearing age.*

C. EVOLUTION OF OCULAR COMPONENTS

Our eyes are fascinating examples of a potpourri of components seen all along the evolutionary trail. Indeed, our eyes contain components previously developed for other uses. For example, our retinal rhodopsin may have come from an ancient bacteria, who may have used it for photosynthesis. Our rounded corneal curvature originally comes from primitive fish. As noted earlier, in the world under the sea, the corneal refractive power is canceled by the surrounding water. However,

---

* Darwinian natural selection only addresses the biologic events of an organism up to the age of reproduction. Thus, in Darwinian terms it is difficult to understand the useful changes of the aging eye.

rounded shape helps decrease water resistance (i.e., be more streamlined). Finally, and most difficult to explain, many of the special crystalline proteins that have been identified in animal eye lenses are similar to metabolic enzymes found elsewhere in the body. For example, lactic dehydrogenase β is similar to E-crystalline found in the lens; arginosuccinate lyase is similar to γ-crystalline found in the lens.[35] The finding that nonlenticular enzymes have been adapted to also function as the special lens proteins, is known as *gene sharing*. Unhappily, the term gene sharing simply names the phenomenon without giving us an idea of which function came first, or what evolutionary pressures forced the new use of the molecules. In summary, one might think of this entire process of nature reaching into its dusty attic to find new uses for old creations as the ultimate in recycling efficiency.

REFERENCES

1. Miller D, Benedek G: *Intraocular Light Scattering.* Charles C Thomas, Springfield, IL, 1973.
2. Miles S: *Underwater Medicine.* Jeppesen Sanderson Inc., Philadelphia, 1966, p. 156.
3. Miller D: *Optics and Refraction, A User Friendly Guide.* Gower Medical Publishers/Mosby, St. Louis, 1991.
4. Aggurwala KR, Nowbotsing S, Kruger PB: Accommodation to monochromatic and whitelight targets. *Invest. Ophthalmol. Vis. Sci.* 36:2695, 1995.
5. Stryer L: Mini review: Visual excitation and recovery. *J. Biol. Chem.* 266(No. 17):1071, 1991.
6. Nakanishi K: Why II-as-Retinal. *Am. Zool.* 31:479–489, 1991.
7. Oesterhelt D, Stoekenius W: Rhodopsin-like protein from the membrane of Halobacterium Halobium. *Nature New Biol.* 1971, 233:149–152, 1971.
8. Spudich JL, Bogomolni RA: Sensory rhodopsins of halobacteria. *Ann. Rev. Biophys. Biophys. Chem.* 17:193, 1988.
9. Atkins GH, Blanchard D, Lemaire H, et al: Picosecond time resolved fluorescence spectroscopy of K-590 in the Bacteriorhodopsin Photocycle. *Biophys. J.* 55:263–274, 1989.
10. Findlay JBC: The biosynthetic, functional and evolutionary implications of the structure of rhodopsin. In *The Molecular Mechanism of Photoreception* (Dahlem Workshop Reports, Life Sciences Research Report 34) (Stieve H, ed.). Springer-Verlag, Berlin, 1986, pp. 11–30.
11. Yokoyama S, Yokoyana R: Molecular evolution of human visual pigment genes. *Mol. Biol. Evol.* 6(2):186–197, 1989.
12. Campbell FW, Gubisch RW: Optical quality of the human eye. *J. Physiol.* 186:558–578, 1966.
13. Fein A, Szutz EZ: *Photoreceptors: Their Role in Vision.* Cambridge University Press, Cambridge, England, 1982.
14. Snyder AW, Bossomaier JR, Huges A: Optical image quality and the cone mosaic. *Science* 231:499–501, 1986.
15. Synder AW, Menzal R: *Photoreceptor Optics.* Springer-Verlag, Berlin, 1975.
16. Ratliff F: The role of physiologic nystagmus in monocular acuity. *J. Exp. Psychol.* 43:163, 1952.
17. Riggs LA: Visual acuity. In *Vision and Visual Perception* (Graham CH, ed.). John Wiley & Sons, Inc., New York, 1965, p. 321.
18. Riggs LA, Ratliff F, Cornsweet JC, Cornsweet TN: The disappearance of steadily fixated visual test objects. *J. Opt. Soc. Am.* 43:495, 1953.
19. Eakin RM: Evolution of photoreceptors. In *Evolutionary Biology,* Vol. 2 (Robzhansky T, Hecht MK, Steere WC, eds.). Appleton-Century-Crofts, New York, 1968, pp. 194–242.
20. Williams DR: Topography of the foveal cone mosaic in the living human eye. *Vis. Res.* 28:433–454, 1988.

21. Von Ditfurther H: *Children of the Universe.* Atheneum Press, New York, 1976.
22. Zeilik M: *Astronomy: The Evolving Universe,* 3rd Ed. Harper and Row, New York, 1982.
23. Thisbos LN, Zhang X, Ming Y: The chromatic eye: A new model of ocular chromatic aberration. In *Ophthalmic and Visual Optics,* Feb 6–8, 1991, Opt. Soc. Amer., Washington, D.C.
24. Aggarwala KR, Kruger PB, Mathews S: Accommodation to monochromatic and white light targets. *Invest. Ophthalmol. Vis. Sci.* 36(13):2695, 1995.
25. Schachar D, Black TD, Kash RL, Cudmore DP, Schanzlin DJ: The mechanism of accommodation and presbyopia in the primate. *Ann. Ophthalmol.* 27(2):58–67, 1995.
26. Fernald RD: Vision and behavior in an African cichild fish. *Am. Scientist* 72:58, Jan-Feb 1984.
27. McPhee J: *The Ransom of Russian Art.* Farvar, Straus, Giroux, New York, 1995.
28. Barlow HB: Critical limiting factors in the design of the eye and visual cortex: The Ferrier Lecture 1980. *Proc. R. Soc (London)* B:212:1–34, 1981.
29. Campbell F: Resolution of the eye and visual cortex. The Ferrier Lecture 1980. *Proc. R. Soc. (London)* B:212:1–34, 1981.
30. Katz M: The human eye as an optical system. In *Clinical Ophthalmology,* Vol. 1 (Duane TD, ed.). Harper and Row, Philadelphia, 1990, Ch. 33.
31. Miller D, Stern R: Vision and hearing screening in the elderly. *EENT Monthly* 13:128–133. 1974.
32. Pizzarellio LD: The dimensions of the problems of eye disease among the elderly. *Ophthalmology* 94:1191–1195, 1987.
33. Weiter JJ: Phototoxic changes in the retina. In *Clinical Light Damage to the Eye* (Miller D. ed.). Springer-Verlag, New York, 1987, pp. 79–127.
34. Hemenger RP, Garner LF, Ooi CS: Change with age of the refractive index gradient of the human ocular lens. *Invest. Ophthalmol. Vis. Sci.* 36:703, 1995.
35. Wistow GJ, Piatigorsky J: Recruitment of enzymes as lens structural proteins. *Science* 236:1554, 1987.
36. Enoch JM: Retinal receptor orientation and the role of fiber optics in vision. *Am. J. Optom.* 49:455–470, 1972.

3

EYES OF DIFFERENT ANIMALS

Introduction
1. The High-Altitude Niche: The Eagle
2. The Tree Niche: The Monkey
3. The High Off the Ground Niche: Tall Animals
4. The Very Close to the Ground Niche: Insects
5. The Ocean Niche: The Sculpin
6. The Air or Land and Water Niche: Diving Birds
7. The Nighttime Niche: The Cat
Summary
    A. Extremes of Optical Performance
    B. The Master Regulatory Gene for Eyes

INTRODUCTION

So far, we have seen how the crude visual system of the infant inches its way to the higher resolution system of the adult human. As noted earlier, the normal human adult with 20/20 vision can resolve two stars separated by only one minute of arc. Is nature capable of producing a better optical system? Might there be animals or birds with more advanced systems? In order to place our ocular optical system in perspective, let us look at those systems that perform at the extreme limits.

During the process of evolution, the fittest creatures dominated a certain geographic area, or a certain time interval within the 24-hour cycle. Sometimes, such a place might be above ground in the trees, or above the trees in the air. These multidimensional spaces that represent the conditions under which an organism best survives are called *niches.* It is in these unusual niches that we will find extreme optical performance.

1. THE HIGH-ALTITUDE NICHE: THE EAGLE

Of all the visual systems of the birds that we might analyze, that of the eagle has the highest optical resolution.[1,2] In fact, on my last visit to my friend and colleague, Robert Stegmann, who directs an eye clinic in South Africa, I actually met a Bateleur eagle face to face. When I announced that one of my lifelong ambitions was to examine the eyes of an eagle, Robert promised to see if something might be arranged. Imagine my surprise when two zoo attendants appeared in his eye clinic carrying a Bateleur eagle, discreetly hidden in a portable cage that resembled a porous suitcase. They were ushered into a special testing room, which was free of patients at that time. Within minutes, the attendants tranquilized the eagle with a device similar to a dart gun, and then intubated the creature and administered a general anesthetic (Figure 3.1 in color insert). When the eagle was asleep, the attendants propped it up in a photographic biomicroscope, and we were able to look at the anterior components of its eye.[3]

What struck us as being the most unexpected feature was the crystal clear cornea. As opposed to the mild light scattering from the fibers and keratocytes distributed within the human cornea, the eagle's fibers scattered no light and had almost no cells. As a matter of fact, because of the absence of backscattered light from its cornea, we could not get a decent photo of the corneal cross-section, no matter how intense a flash we used (Figure 3.2). The second feature that surprised us was that its pupils did not get smaller as the bright light was shown in its eye

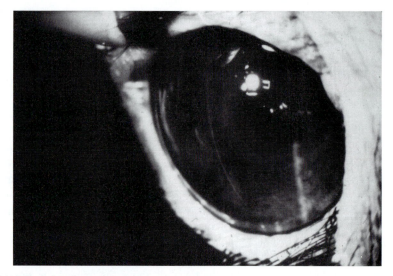

FIGURE 3.2   Photo of the cornea of the batteleaur eagle showing very little corneal detail because of the paucity of backscattering elements in the corneal stroma, even though we used the most sensitive black and white film. (From Miller D, Stegmann R: The eye of the eagle. *Eur. J. Implant. Refract. Surg.* Vol. 3, March 1991, with permission.)

(Figure 3.3 in color insert). Apparently, the bird functions with a relatively large pupil much of the time. In the human, a large pupil invites image degradation from spherical aberration. Not so in the eagle eye. Experiments done on captured eagles that have been trained to be laboratory subjects, show that they can resolve objects at least four times smaller than we can in bright light.[4,5] Such a high degree of resolution could not take place if spherical aberration was present in their eyes.

There were other features that surprised and amazed us. The sclera was black (Figure 3.4 in color insert), presumably to prevent bright glaring light from leaking onto the retinal image from the side. Within the vitreous body of the eye was a black shadelike structure that works like an internal visor. This structure is called the *pecten* (Figure 3.5) and probably protects the fovea from the sun when it is below the eagle. The configuration of the fovea was that of a deep pit (Figure 3.6). Hence, the fovea was protected from stray light by the steep walls of the pit.[4] Since birds have very little brain structure positioned between their eyes, the axial length of each eye can be measured by measuring the distance from the apex of one cornea to the other (Figure 3.7) and dividing by 2. When you do that, you find

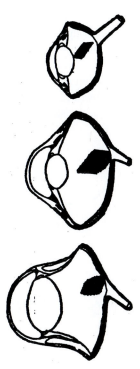

FIGURE 3.5 Diagram of the pecten in three different birds' eyes. It is a coil of blood vessels (which nourish the retina), is covered with black pigment, and is found in the vitreous cavity. (From Burton M: *The Sixth Sense of Animals.* Taplinger Publishing, New York, 1972, p. 125, with permission.)

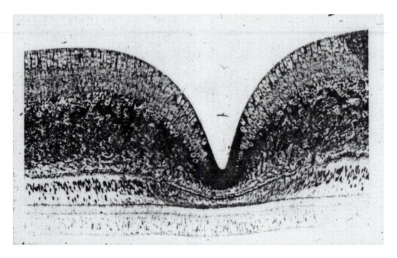

FIGURE 3.6 Histologic specimen of the eagle's fovea. Note the deep pit configuration, which protects the foveal receptors at the bottom of the pit from glare. (From Regmond L: Spatial visual acuity of the eagle, Aquila audax: a behavioral, optical and anatomic investigation. *Vis. Res.* 25:1477–1491, 1985, with permission.)

the axial length to be about 35 mm long (the human eye averages 24 mm in length). We can now start to appreciate how the eagle does so well in its airborne niche by analyzing its visual system.

Without the benefit of trees or ground cover, the sun's glare would present a major problem to the human eye. However, the black sclera, the black visorlike pectin, and the fovea at the bottom of the steep-walled fovea all present sizable

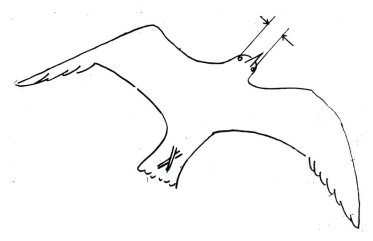

FIGURE 3.7 Axial length of an eye is measured in a bird by first measuring the distance from one cornea to the other and dividing by 2.

obstacles to the degrading effect of glare in the eagle eye. Stalking small animal prey a few thousand feet above the earth's surface might be an impossible task for a human without the aid of binoculars. However, the eagle eye produces a large retinal image, because of its longer axial length. You might think of the effect of the longer eye being somewhat like that of the long bellows system on the camera used by professional photographers. The eagle is also able to extract more information from the retinal image by three optical mechanisms. Its optical system suffers from less inherent image blurring because of (1) the lower levels of light scattering in the cornea, (2) less spherical aberration overall, and (3) less diffraction because of its large pupil. In addition, its retina behaves more like a fine-grain, high-resolution film because the size of its foveal cones are finer than those of the human. Finally, since the blood supply to the retina comes from only the pecten and the choroid,[6] there are no blood vessel nets (humans have two such nets) within the retina to cover any of the photo receptors, and produce angioscotomas.

2. THE TREE NICHE: THE MONKEY

Trees would seem to be a safe home for many animals. In reality, a fight for survival is present within the tree environment between insects, reptiles, amphibians, birds, and a few mammals. In this section, we focus on a spectacular mammal that has found a relatively safe haven within the trees, the monkey.[7,8]

Because of the sunny nature of life near the tree tops, glare is a problem for high-climbing monkeys. Although the monkey does not possess all the antiglare devices of the eagle, many have black sclera (Figure 3.8 in color insert). The virtue of black sclera was made clear to me a few years ago by an experience I had with a patient. A little girl with albinism was brought to our clinic. Such a patient has very little dark pigment. Consequently, her hair was blonde in color, and her skin quite pale. The lack of pigment affects albino eyes in a most distinct manner. The choroid layer, which normally is loaded with dark, melanin pigment, and functions as the light-absorbing layer lining the sclera, has no dark pigment in the albino patient. Thus, these patients are extremely glare sensitive. To make matters worse, the iris also lacks a pigment backing. Hence, stray light is able to come through the sclera and iris and severely degrade the retinal image because of the glare. If that weren't enough, the foveal cones in these patients are also poorly formed. Of course, we have no way, at present, of making better foveal cones. However, we were able to give the little patient better sight by creating a pair of dark glasses to compensate for the pale iris, and dark side shields to compensate for the light leak through her sclerae (Figure 3.9). This patient can be appreciated as representing a somewhat parallel situation to the monkey high in the trees. Both, for different reasons, are bothered by large amounts of side-directed glaring light. Both are better able to cope with side glare by either a black sclera or a manmade equivalent.

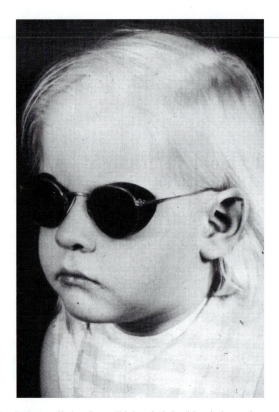

FIGURE 3.9 Patient suffering from albinism is helped by dark sunglasses with side shields (From Miller D, Farley VH, McLaughlin R, Sullivan GE: A light shielded spectacle for albino patients. *Ann. Ophthalmol.* Aug. 1972, p. 612, with permission.)

The monkey must also be able to recognize strangers and friends on the ground. Although its resolving system is not quite as good as the eagle's, some monkeys have a system that is similar to the human's. If you have ever looked out of a window from the sixth floor of an apartment building, you know that you can accurately recognize important objects on the ground. The monkey has a similar level of visual resolution from its tree home.

The last monkey visual trait that suits its niche so well is its stereoscopic vision. Clearly a refined sense of depth is helpful in leaping from branch to branch. By placing its two eyes in front of its head and devoting a large area of visual cortex to the processing of three-dimensional information, the monkey can access any part of the tree system from base to canopy by its nimble leaps. The monkey's stereopsis also allows it to better detect a well-camouflaged snake, which may blend with its surroundings, but which sticks out, spatially, from its surroundings.

3. THE HIGH OFF THE GROUND NICHE:
TALL ANIMALS

As land mammals have evolved, certain species have increased in body size. The enlarging herbivores needed to graze greater tracks of land. By the same token, the enlarging carnivores expanded their territories in pursuing their prey. Interestingly, there is a relation between body size and eye size. It is not a simple linear relationship. The logarithm of the retinal magnification factor (which is proportional to the axial length of the eye) does vary almost linearly with the logarithm of the body weight.[9] The need to use logarithms becomes clear as the axial length of eye goes from 0.8 mm in the tiny tree shrew to 54 mm in the baleen whale, while the weight of the shrew is almost 100 g, compared to the many tons of the whale. This relationship can be seen graphically in Figure 3.10. The data points actually fall into three main curves. Birds show a steep linear relation suggesting that the retinal image gets rapidly larger as the bird gets heavier. Most of the land animals follow a more gradual slope, with humans and the seal having the largest eyes in this second group. Finally, there is a flat line, hovering about the previously mentioned two curves. It represents tall mammals such as ostriches, zebras, horses, giraffes, and elephants, all having eyes measuring about 50 mm (twice that of the human).

Let us look at an analogous situation in cameras. To get a good close-up or portrait picture, you have to get very close to your subject with a traditional small

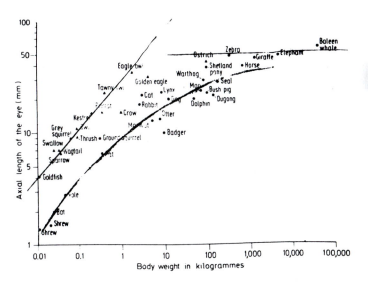

FIGURE 3.10   The relationship between the logarithm of body weight and the retinal magnification factor (which is related to the axial length of the eye). (From Hughes A: The topography of vision in mammals. In *The Visual System in Vertebrates* (Creseitelli F, ed.). Springer-Verlag, Berlin, 1977, with permission.)

There are 3 ways to make the retinal image larger
1. Use a larger object

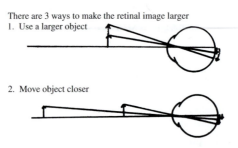

2. Move object closer

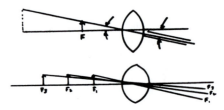

3. We can enlarge the object with a lens
    also put the image at infinity so accommodation will be no problem

FIGURE 3.11   A demonstration of how to get a magnified camera image by moving closer to the subject.

camera or stand at a comfortable distance across the room and add a special lens or an extension bellows to the camera body. In Figure 3.11, we can see that moving very close to a subject with a close-up lens can produce a large image. In Figure 3.12 (see color insert), the addition of the extension bellows also produces a larger image, for a distant object. For birds and mammals, nature creates a large retinal image of a distant object by using the bellows system. In essence, the bellows gives the camera a greater axial length. An eye with a longer axial length also creates a larger image. The longer axial length can be thought of as a way to magnify images of distant objects. It is a tool for a creature that must identify prey or a predator in the distance.[10]

As noted, tall animals have an eye twice the size of the human. We know that most humans range from 4 to 6 feet in height. Creatures such as horses, zebras, and ostriches stand 8 to 10 feet high, while giraffes may be 12 feet or taller. If their eyes were the size of the human, the retinal image of a pebble, seed, or ground insect would be half the size of that seen by the human, whose eyes are closer to the ground. Hence, the longer eye provides larger retinal images for both close and distant objects. Two benefits are possible if these mammals had the equivalent of the primate or bird foveae. First, their eyes can be considered to be 2× telescopes. This would suggest that their resolution would be twice as good as the primate eyes. Since experiments suggest that their resolution is not better than ours, we must look for another possible explanation. If a fovea densely packed with fine cones is found in only birds and primates, then we may ask if there is

FIGURE 3.1 Intubation of batteleaur eagle in Dr. Stegmann's eye clinic. (From Miller D, Stegmann R: The eye of the eagle. *Eur. J. Implant. Refract. Surg.* Vol. 3, March 1991, with permission.)

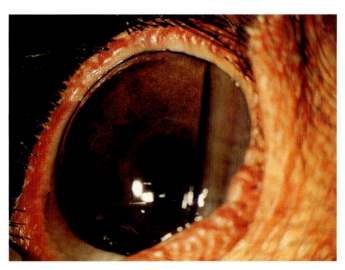

FIGURE 3.3 Moderately dilated pupil of the batteleaur eagle despite the presence of bright light. Note the tan nictating membrane covering the bottom of the cornea.

FIGURE 3.4   The bald eagle eye demonstrating a black sclera, which probably diminishes glare from side lighting. (From Mackenzie JPS: *Birds of Prey.* Northword, Minosqua, WI, 1946, p. 106, with permission.)

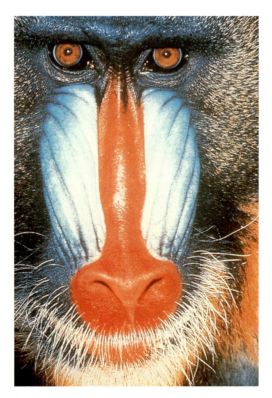

FIGURE 3.8   Note the black sclera of this mandril monkey. (From Martin M, May J, Taylor R: *Weird and Wonderful Wild Life.* Chronicle Books, San Francisco, 1983, cover photo, with permission.)

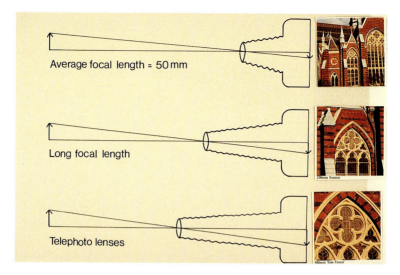

FIGURE 3.12 A demonstration of how to get a magnified camera image by using an extension bellows.

FIGURE 3.13 The long eyelashes of the giraffe act as a visor to prevent glare (From Ross K: *Okarango, Jewel of the Kalahari.* BBC Books, Woodlands, London, 1987, p. 42, with permission.)

FIGURE 3.18 A photograph of an unusual cat mutant. Note that only one eye has a tapedum lucidum. (From *Science*, cover photo August 1987, photographer Cindy Glassauer, with permission.)

FIGURE 3.14 A camera with a large depth of focus captures both this construction worker, high on a building, and the distant objects on the ground. (From National Geographic Society: *The Incredible Machine*, Washington D.C., 1992, photographer Lynn Johnson, Blackstar, p. 29, with permission.)

another way to achieve equivalent resolution. A magnified image from a larger eye focused on a "larger grained" retina should perform almost as well. Hence, the larger image might simply be another way of achieving high resolution in the absence of a fovea.

The tallest animal in this group presents another unique visual feature that is important in this niche. The giraffe is really too tall to be able to station itself under a tree and benefit from the shade. Thus, it needs to be protected from strong glare. To illustrate how potent an effect glare has on vision, I am reminded of a strategy used during World War II. The British were desperately trying to protect the Suez canal from enemy bombing. Since their anti-aircraft defense was no match for the enemy planes, they had to take a creative approach. Could they make the canal invisible? Oddly enough, the military turned to a professional magician named Jasper Maskelyne.[11] He recommended that the use of powerful glaring spot lights along the Suez canal would make the canal disappear. The glare created by the spot lights degraded the pilots' vision to such a degree that the canal actually became invisible to them. The glare of the sun, like the spot lights, can also degrade vision. We are all familiar with the baseball outfielder who loses a fly ball in the sun. To avoid this situation, the ball player wears a peaked cap and flippable sunglasses. As noted in Figure 3.13 (see color insert), the giraffe has its own version of the peaked baseball hat. The excessively long, thick eyelashes help keep away many of the glaring rays of the sun.

4. THE VERY CLOSE TO THE GROUND NICHE: INSECTS

There is a unique natural niche above the ground, yet below the grass and leaves. In this exclusive layer, we find the busy world of insects.* Compared to the vistas of birds or tall animals, the important world of the ground insect goes from almost an antenna length (perhaps $1/4$ inch) to about 12 inches away. Thus the accommodative equivalent range would be from about 3 to 160 diopters. Clearly such a life requires an eye with a large depth of focus. On another level, the insect must also be able to operate in the low lighting found under foliage during the day and at night. Camera buffs will recognize that the optical system of the insect must have an efficient light-gathering system, or a system with a low F number. Low F number systems usually have *large* pupillary apertures. Which type of system could accomplish these tasks in the simplest manner for a small creature? Ironically, the most efficient way to achieve a wide depth of focus is with a *small* aperture system. A familiar example is the pinhole camera system. The old Kodak Brownie had a single lens system, along with a very

---

* Although the oldest insect fossil is probably more than 390 million years old, the ocular optics of today's insects are quite sophisticated in terms of low light sensing. Of course, we do not know if the early insects had such sensitive visual systems.

small pupillary aperture. This meant that a child could simply point and shoot the camera and have almost everything in focus. In a bright outdoor setting with plenty of light, objects as far away as distant mountains, or friends' faces as close as a few feet would all be in sharp focus (Figure 3.14 in color insert). How can the insect have both an enhanced light-gathering system, and a large depth of focus?

The insect eye (Figure 3.15) is known as a compound eye, because it has a large number of tubular optical systems held together much like a bundle of pipes. Each optical tube will focus part of the image onto the insect's retina. Because each tiny tube averages 25 thousandths of a millimeter in diameter, each tube (ommatidium) can be considered to be a very small pupillary aperture system, with a considerable depth of focus. Doesn't a small pupillary aperture limit the amount of light falling on the retina? After all, the human pupil dilates in a dark environment to capture more light. The insect eye uses a second optical principle to maximize its catch of available light.

I discussed earlier about diffraction being responsible for smearing a sharply focused point. I also noted that the diffraction pattern got larger as the pupillary aperture got smaller. Therefore, a very small pupil produces a large focal smear that represents a lower density of photons per retinal area. Such a low density of photons would produce a very poor low light sensing system. However, even in the face of a tiny pupillary aperture, the diffraction pattern can be made much smaller (and the concentration of light much greater) with an optical system that

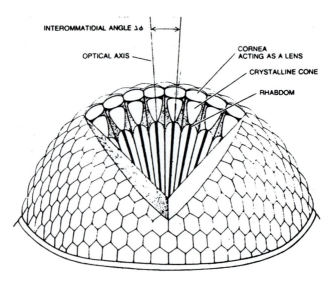

FIGURE 3.15 Diagram of the typical insect eye showing the many tubelike ommatidia. The small apparatus of each ommatidia yields a large depth of focus. (From Horridge GA: *The Compound Eye and Vision of Insects.* Clarendon Press, Oxford, 1975, with permission.)

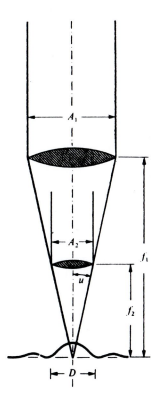

FIGURE 3.16 Diagram showing that the small focal length of an ommatidium works with the small pupillary aperture to efficiently capture photons and produce a system with a low F number (From Lythgoe JN: *The Ecology of Vision.* Oxford University Press, Oxford, 1979, p. 38, with permission.)

is very strong (i.e., possesses a very short focal length).* Indeed, the average focal length for an insect ommatidium is about 50 thousandths of a millimeter.† (Figure 3.16). This combination of small pupil and small focal length gives the insect eye a moderately fuzzy image, but an F number of about 2.[10]

Hence, the insect eye has a large depth of focus, a high F number, and keeps objects in focus at very close range. Such a system is ideal for differentiating friend from foe, movement from stillness, and a few other important pieces of information. However, it is not a high-resolution system, and cannot make out fine details.

* The eye's axial length is related to the eye's focal length. Thus, the very small eyes of the insect have very short focal lengths.

† A lens system with a focal length of 0.050 mm is also said to have a power of 27,300 diopters. This is very strong compared to the power of the human optical system, which averages 60 diopters.

5. THE OCEAN NICHE: THE SCULPIN

The ocean is populated by fish with a fascinating assortment of optical systems. In order to truly appreciate the sophistication of some of these systems, I will choose a fish eye with some very unusual traits. I know the trait is unusual because of a conversation that I had with a well-known British academic ophthalmologist. We both had been invited to address the annual meeting of Pakistani ophthalmologists in Peshawar. The meeting was formally called to order by General Zia, the country's leader at that time. The British professor and I were assigned seats next to each other at the opening ceremony. After we were seated, we were all told that the General would arrive late. Because the General's life was always threatened by enemies, he tried to baffle potential assassins by always confounding his schedule.* To fill in the time, I tried a bit of small talk with my British colleague. I was greeted with a strong silence. Perhaps some politics, maybe some academic gossip. Each well-meaning attempt was greeted with an uncomfortable silence. We had been seated for about an hour, and I was coming closer to my frustration threshold. Well, one more try, I thought. "We're doing some interesting research on an unusual fish eye in a lab in Boston these days." No response. "Yes, the creature's cornea turns bright orange when we expose it to bright light." "Really," he responded, "How does it accomplish this?" Obviously, I had pressed the right key. "Well, the limbal area of the cornea has many bulging flask-shaped cells, whose long slender cell processes lead to the corneal center.[12] When we expose the fish to bright light, the pigment concentrated in the bulging flask part of the cell migrates along the slender processes toward the center, producing an orange brush-stroked appearance." Indeed, the fish, known as the long horn sculpin, has a clear cornea in deep water, but converts to an orange cornea when it rises to the surface of the ocean where the light is brighter. This particular trait is actually seen in a number of fish species. Since those species living at the greatest depths have the most intense coloring when rising to the surface, it is felt that the coloration functions to prevent light damage to the retina.[12] As opposed to our eyes, where the retina adapts when going from a dark setting to one of brightness, the sculpin probably maintains a constant adaptation in its retina by putting on its "photochromatic sunglasses."

The long horn sculpin possesses a second corneal feature that is unusual. If its cornea is scratched, the injured area heals five times faster than the human cornea.[13] Dr. Hank Edelhauser, who noted the phenomenon along with his students, said that he could almost see the new cells moving into the injured area. I assume that the sculpin hides in dark parts of the ocean where plant and rock obstacles are ubiquitous. The price it must pay for such a protective home are frequent corneal abrasions. By healing such abrasions in a few hours, it can prevent its cornea from becoming waterlogged and cloudy (scatters more light as spaces

---

* Although General Zia's security system was tight, he was ultimately assassinated in a helicopter explosion.

between collagen fibers enlarge). Hence, it can successfully survive in its unusual niche in large part due to its agile repair system.

6. THE AIR OR LAND AND WATER NICHE: DIVING BIRDS

Diving birds fall into this niche. For example, the cormorant flies over the surface of the water looking for fish. Although it looks quite simple, the cormorant must have a very special optical system in order to see clearly in the air and then under water. In air, the cormorant's cornea, just as the human's, is a powerful focusing element. Corneal focusing power depends on the large difference in index of refraction between air and the tissue of the cornea. But, how are they to keep their prey in focus under water, where the power of the cornea is neutralized? The answer is that they have an extraordinary range of accommodation. In Figure 3.17, we can see how the lens inside the eye of another diving bird, the hooded merganser, changes shape in order to make up for the lost corneal power and then some. These birds show range of accommodation of almost 80 diopters.[14,15] This

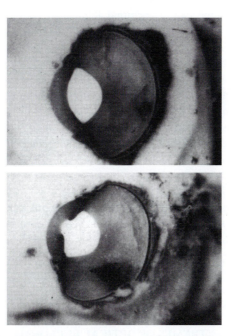

FIGURE 3.17 An experiment on the enucleated eye of the hooded merganser in which the eye lens is made to change shape by electrically stimulating the eye. Note how the radius of curvature of the eye lens changes dramatically upon stimulation (From Levy B, Sivak JG: Mechanism of accommodation in the bird eye. *J. Comp. Physiol.* 137:267–272, 1980, with permission.)[17]

is quite an impressive performance when we compare it to the accommodative amplitude of the average young human adult, which is about 10 diopters.

The penguin can be considered a walking, rather than a flying, diving bird. It too must function in air and under water. Its ocular modus operandi is similar to that of the hooded merganser, but with a variation. Its cornea, while on land (i.e., surrounded by air) is quite flat and has very little optical power. For example, the Humboldt penguin has a 10 diopter cornea, and the Rockhopper penguin has an 11 diopter cornea.[16] Compare this figure to the human with an average corneal power of about 42 diopters. Hence, the penguin eye gets little contribution to focusing power from its cornea even on land. This fact tells us that certain penguins must have a large amplitude of accommodation simply to operate in its land niche. Under water, its 10 diopter cornea becomes neutralized, telling us that its accommodative apparatus must work even harder to keep its prey in focus.

7. THE NIGHTTIME NICHE: THE CAT

So far we have discussed evolutionary niches as layers in a geologic birthday cake. Evolution has not only created place niches, but also time niches. For example, some animals function during daylight hours, while others work only at night. As you can imagine, the eyes of nocturnal animals are packed with low-light level amplifiers. The lowest F number (an index of light-gathering power dependent on the ratio of focal length to pupil size) belongs to the net casting spider at a record of 0.58. Not far behind is the cat with an F number of 0.89, followed by insects at about 2, and the dark adapted human at over 3.[10] However, the cat has a second device (used by many animals) that improves its photon catching ability even further. Most of us can recall driving at night with our headlights turned on. Suddenly, along the side of the road we will see the eye sheen of an animal, such as a dog or a cat. The sheen is due to a reflecting layer of tissue just behind the retina called the *tapedum lucidum.* Any light that is not initially absorbed by the retinal pigment gets reflected back by the tapedum lucidum, offering the retina a second chance to register the light. In Figure 3.18 (see color insert), we can see the eyes of an unusual cat.[18] One eye has a reflecting tapedum, while the other eye does not. When researchers studied electrical recordings from each retina under very low lighting conditions, they found that the eye with the tapedum showed almost twice the light sensitivity as compared to the eye without the tapedum.

SUMMARY

A. EXTREMES OF OPTICAL PERFORMANCE

In evaluating the optical system of birds like the eagle, one is struck by the extreme level of performance. Usually, the pupil of the eagle eye stays enlarged even in bright light. Camera buffs know that cameras capable of sharp imagery using large pupillary apertures have complicated optical systems in order to compensate for spherical and chromatic aberration. The eagle eye is capable of a resolution probably four times better than that of the human eye, limited only by diffraction, an irreducible phenomena of the laws of physics. As noted above, the bird's pupil remains large in bright light, which diminishes the degrading effect of diffraction. A biologic optical system that functions at the outer limits of the laws of physics is indeed a wonder.

Not mentioned in this chapter is the large field of view of all the animals and birds that have their eyes placed at the side of the head, giving some almost a full 360° field of view (most humans have about 180° field of view).

B. THE MASTER REGULATORY GENE FOR EYES

A discussion of different eyes in different animals living in different niches makes one question whether eyes evolved linearly from an appropriate branch, when a special niche developed.[19] Certainly, the simple eyes of vertebrate resemble each other enough that the idea of all the eyes stemming from a primordial eye seems tenable. Thus, we might assume that our eyes come from this relatively recent evolutionary branch. The compound eyes of most insects and anthropodia (Figure 15) are constructed so differently and are so much smaller, that surely they must have evolved from a unique beginning with their own branching tree of mutations. Therefore, it was not too surprising when a master regulatory gene was isolated in the fruit fly, which was shown to be responsible for the basic organization of its eye. The gene was then introduced into different parts of the fly's body and produced typical fruit fly eyes on the feet, trunk, etc.[20] However, what really surprised everyone was that this same gene was found in both mice and humans.[21,22] In fact, if the gene was removed from the developing mouse, it would be eyeless. The presence of this common organizational or design gene suggests that an elemental plan to have two eyes attached to the brain (with the molecular components found in eyes of all species) appears very early in the course of evolution, before there was much branching into the different species. The presence of these genes further suggests that this plan for a primordial visual system proved so effective for survival that it remained unchanged through the eons of time underlying evolution. It is quite remarkable that the basic organization of the visual system, whose common components remain, but can be put together in different ways, has survived for hundreds of millions of years.

REFERENCES

1. Downer J: *Super Sense: Perception in the Animal World.* Henry Hoet & Company, New York, 1988.
2. Meyer DB: The avian eye and its adaptation. *Handbook of Sensory Physiology,* Vol. VII/5th Ed. (Crescitelli F, ed.). Springer-Verlag, Berlin, 1977.
3. Miller D, Stegmann R: The eye of the eagle. *Eur. J. Implant Refract Surg.* Vol 3, March 1991.

4. Reymond L: Spatial visual acuity of the eagle: A behavioral, optical and anatomic investigation. *Vis. Res.* 25:1477–1491, 1985.

5. Sinclair S: *How Animals See.* Facts on File Publication, New York, 1985.

6. Chase J: The evolution of retinal vascularization in mammals: A comparison of vascular and avascular retinae. *Ophthalmology* 89:1518, 1982.

7. Wald G: The distribution and evolution of visual systems. In *Comparative Biochemistry,* Vol. 1, (Florkin M, Mason HS, eds.). Academic Press, New York, pp. 311–345, 1968.

8. Walls GI: *The Vertebrate Eye and its Adaptive Radiation.* Bloomfield Hills, MO: The Cranbrook Institute of Science, 1942.

9. Hughes A: The topography of vision in mammals. In *The Visual System in Vertebrates,* (Crescitelli F, ed.). Springer-Verlag, Berlin, 1977, p. 654.

10. Lythgoe JN: *The Ecology of Vision.* Clarendon Press, Oxford, 1979.

11. Fisher D: *The War Magician.* Coward-McCann Inc., New York, 1983.

12. Orlov O, Yu, Camburtseva AG: Changeable coloration of corneas in the fish hexagrammos octogrammus. *Nature* 263:405, 1976.

13. Ubels, JL, Edelhauser, HF: *Healing of corneal epithelial wounds in marine and fresh water fish. Curr. Eye Res.* 2:613–619, 1983.

14. Sivak JG: Accommodation in vertebrates: A contemporary survey. In *Current Topics in Eye Research,* Vol. 3, (Zadunaisky JA, Davson H, eds.). Academic Press, New York, 1980, pp. 281–330.

15. Sivak JG: Optics of the crystalline lens. *Am. J. Optom. Physiol. Optics* 62:299–308, 1985.

16. Sivak JG: Corneal optics in squatic animals: How they see above and below. In *The Cornea: Transactions of the World Congress on Cornea III* (Cavanagh HD, ed.). Raven Press, New York, Chap. 32, 1988 (diving birds).

17. Levy G, Sivak JG: Mechanisms of accommodation in the bird eye. *J. Comp. Physiol.* 37:267–272, 1980.

18. Pion PD: Myocardial failure in cats associated with low plasma taurine. *Science* 237:764, 1987.

19. V. Salvini-Plawen L, Mayr E: On the evolution of photoreceptors and eyes. *Evol. Biol.* 10:207–263, 1997.

20. Halder GP, Callaerts, Gehring WJ: Induction of ectopic eyes by targeted expression of the eyeless gene in Drosophilia. *Science* 267:1788–1792, 1995.

21. Graham AN, Papalopulu, Krumlauf R: The murine and Drosophilia homeobox gene complexes have common features of organization and expression. *Cell* 57:367–378, 1989.

22. Quiring R, Walldorf U, Kloter U, Gehring WJ: Homology of the eyeless gene of Drosophila to the small eye gene in mice and Aniridia in humans. *Science* 265: 785–789, 1994.

4

THE HEALING EYE

Introduction
1. Corneal Abrasion
2. Corneal Laceration
3. Corneal Laceration and Traumatic Cataract in a Child
4. Cataract and Iridotomy from Blunt Trauma
5. Blowout Orbital Fracture
6. Blunt Trauma and the Oculo-Cardiac Reflex
7. The Other Eye
Summary
    A. Natural Eye Repair and the Community
    B. Eye Healing in Certain Animals

INTRODUCTION

The front surface of the eye constitutes only 0.27% of the body surface, and occupies 4% of the area of the face. Yet eye injuries comprise up to 10% of all bodily injuries. Why does such a small area seem to attract such a disproportionately large number of injuries? Professor Michael Belkin of the Sackler School of Medicine at Tel Aviv University speculates that trauma caused by a tiny flying object is hardly noticed when it strikes the skin. If such a missile were to directly hit the eye, the result is usually a sudden decrease in vision and severe pain. What is even more frightening is that on rare occasions, a badly injured eye can cause blindness in the other eye, due to an unusual autoimmune reaction known as *sympathetic ophthalmia*.

Clearly, the eyes seem to receive an unfair share of injuries. The National Society to Prevent Blindness has estimated that more than 2.4 million eye injuries occur each year in the United States.[1] That is equivalent to saying that almost 1% of the American population gets an eye injury each year. By extending that figure, we might conclude that almost 20% of the population will sustain an eye injury in the next 20 years (assuming no one gets injured more than once). My medical experience in the underdeveloped parts of the world, where eye safety is not promoted as it is in the United States, suggests that the incidence of eye injury in those countries might be double or triple that number. This means that in 20 years, over half the members of a tribal community might sustain an eye injury. How many of these injuries are serious enough to lead to blindness if not treated properly? We can divide eye injuries into two groups. In the United States about 10% of eye injuries are serious enough to require hospitalization. The second group (90%) are scratches to the front of the eye that heal quickly.[2] In underdeveloped countries, many of the patients with injuries (caused by flying particles from digging and chopping, or from stones, fists, branches, or thorns) never reach a hospital and many of the wounds become infected, ultimately leading to a corneal scar and a significant loss of sight in that eye.

Can we use this information to understand the conditions of our primitive ancestors? If we use this information then a large percentage of our primitive ancestors sustained eye injuries and became partially blind in one or both eyes. A significant percentage of blind people would play a prominent role in the evolution of human societies. But if the incidence of eye injuries in the people of the underdeveloped world are similar to our human ancestors, then something is wrong. United Nations statistics indicate that the prevalence of true blindness in the underdeveloped world from all causes is between 1% and 2%. A study evaluating the data from Bedouin eye camps suggests that eye injuries accounted for only 10% of all cases of blindness.[3] Therefore, eye injuries probably produced true blindness in only a small percentage of our ancestors, although many may have suffered from some visual impairment.

How can we explain such a high injury rate associated with such a low blindness rate? The answer is that the reparative system of the eye is very effective. I would like to showcase here the way the body heals some typical eye injuries.

1. CORNEAL ABRASION

A number of years ago, I worked in Cartegena, Colombia. On one of my afternoons off, I was invited to a bullfight. During one of the encounters, I noticed that the matador's eye had been grazed by a swirling banderilla (an arrowlike device with feathers that was stuck in the bull's back). The stunned matador stopped in his tracks, dropped his cape, and simply held his eye, unable to function. Fortunately, an alert colleague pulled him out of the ring before the bull noticed the frozen bullfighter.

The matador was quickly attended by a doctor, who I guessed, placed a drop of anesthetic in the bullfighter's eye. With the pain gone, the matador returned to the ring feeling much better. He killed the bull in one pass and quickly left the ring without the usual bowing before the cheering crowd. Because my host at the fight knew the doctor covering the bullfight, I was introduced to the doctor and invited to examine the matador. My assumption was correct; he had suffered a scratched cornea from the swirling banderilla. As the local anesthetic wore off, his eye became painful again and began to tear profusely. He could see very little with the eye. When I saw him again a week later, his vision was perfect, and his wide smile told me that the pain was gone.

Let us look at a typical corneal abrasion in greater detail (Figure 4.1 in color insert). We know from histology studies in animals that a scratch on the corneal surface looks like a divot on the golf course. The surface cells have been scraped away, leaving a small crater. Since the corneal surface has the densest supply of nerves in the body, irritation of these nerves is responsible for the intense pain.

Upon injury, the lacrimal gland immediately washes the injured cornea with an outpouring of tears. This profuse tearing works to blur the vision, much like looking through a waterfall. In addition, the blood vessels of the conjunctiva become engorged and leak serum and white blood cells, which will tend to disinfect the scratch.

As expected, the corneal scratch interferes with vision much like a deep scratch in your spectacles would interfere with sharp focus. However, the most fascinating aspect of this injury is the pattern of the healing of the epithelial cells.

The original smooth corneal surface that was optically perfect is restored by the healing corneal epithelial cells, which fill in the scratch and re-create the original corneal curvature (Figure 4.2 in color insert). In fact, if the scraping injury has been very extensive and most of the corneal epithelial cells have been removed, the limbal stem cells of the cornea will migrate over to cover the cornea, slowly lose their limbal character, and take on all the clear optical quality of the central corneal epithelium, ultimately reproducing the original corneal curvature.

In emergency departments in the United States, we usually treat such corneal abrasions with local antibiotics to prevent infection and oral anesthetics and antiinflammatory agents to dull the pain. However, if left to their own natural devices, most abrasions will heal by themselves in a few days.

2. CORNEAL LACERATION

Much of my thinking about eye injuries was developed working in Dr. Robert Stegmann's clinic in South Africa. In the 1980's we had developed the use of the viscoelastic agent Healon for intraocular surgery. Once the techniques were refined, he used the jelly to successfully repair a large number of badly traumatized eyes, often considered hopeless in his clinic. As his reputation grew, a local philanthropist donated a special camera for his operating room. With the camera, he would record all the difficult surgical repairs. When I would visit him in Africa, we would review these filmed cases. For example, we reviewed the case of a young man whose eye had been pierced by a recoiling wire, as he carelessly opened a large carton. The man had been brought to the clinic practically blind in that eye, in severe pain, and with profuse tearing. The wire had slashed open the cornea. In response to the laceration, the edges of the wound had swelled to more than double their normal thickness. A portion of brown iris had prolapsed and was glued stuck in the wound, preventing the internal ocular fluid from leaking out, much as the little Dutch boy, Hans Brinker, had plugged the opening in the dike with his finger. The prolapsed iris became glued in place by fibrin, and required an arduous dissection in order to free the iris and place it back into the eye. Once the iris was surgically freed and replaced into its natural position inside the eye, the large wound was sutured closed. After about 3 months, the patient could see more clearly (i.e., 20/40 in that eye).[4]

As we repeatedly watched the operation on film, the scenario of steps that the body took to contain the damage and work toward a level of useful vision, after healing, became clearer. The surgical philosophy of the day was that surgical repair of an injured eye was difficult because of all the road blocks (i.e., adhesions) that the healing process created. However, we started to look at the eye's responses from the vantage point of the eye's own survival. Clearly, the eye's priorities seemed to be to first plug the fluid leak and prevent potential infection. In a second stage, it was to guide the healing so as to restore some useful vision. Of course, modern surgical repair techniques improve the visual outcome better than natural healing. Nevertheless, we felt it important to review the sequence of body responses in order to learn from a strategy that had taken millions of years to evolve. For example, we knew that the edges of the corneal wound swelled because the damaged cells at the back of the cornea allowed extra fluid to penetrate and waterlog the tissue (Figure 4.3 in color insert). We also knew that the elastic iris came forward at the time of the injury, because the pressure within the eye exceeded the atmospheric pressure outside the eye (Figure 4.4 in color insert). We speculated that the swollen edges of the wound produced a larger surface area and so offered greater friction and held the iris plug more firmly in place. Later, the iris secreted fibrin that covered its surface and literally glued itself to the cornea. Ironically, it appeared as if the wound is able to set off a cascade of biologic events at the front of the eye that can ultimately swell and detach the choroid layer at the back of the eye. When this happened, the vitreous body is pushed forward. By coming forward, it forces all the tissues in front of it forward, thus maintaining the iris in place while the fibrin glue polymerizes.

After generating these speculations, we understood that if the eye is surgically repaired, then there is no need for many of nature's responses. In fact, some of these natural responses become obstacles to surgical repair (as had been noted). For example, we now often postpone surgery for a day so that we can continually bathe the injured eye with local antibiotics and anti-inflammatory drugs. The local antibiotics work along with the tears to disinfect the wound and prevent infection. The surgical films had taught us that surgery itself can be thought of as a form of trauma to which an irritated eye responds with further swelling and

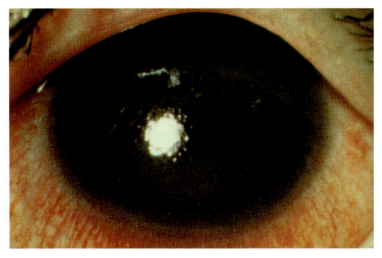

FIGURE 4.1   Clinical picture of corneal abrasion. Note the irregular reflection from the abrasion and the redness on the usually white conjunctiva.

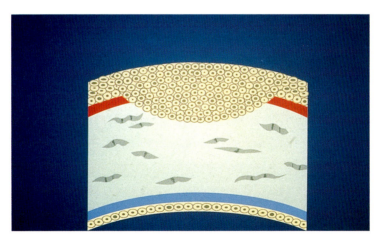

FIGURE 4.2   Drawing of the histology of healed abrasion. Note the way the epithelial cells perfectly fill in the defect. (Courtesy of Roger Steinert, M.D.)

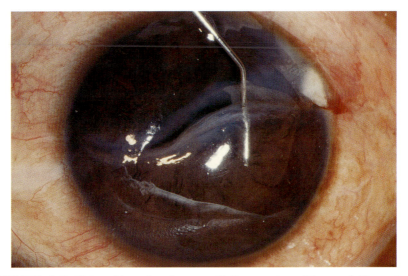

FIGURE 4.3 Picture of laceration of cornea with the gray edges of the wound. The surgical cannula has been placed between the edges of the laceration. The grayness implies a thickening or waterlogging of the edge. (Courtesy of Robert Stegmann, M.D.)

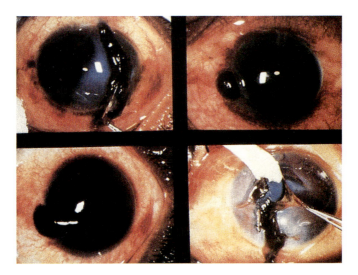

FIGURE 4.4 Four examples of how the elastic iris will prolapse to plug four differently located wounds of the cornea. (From Miller D, Stegmann R: *Treatment of Anterior Segment Ocular Trauma.* Medicopea, Montreal, 1986, cover photo, with permission.)

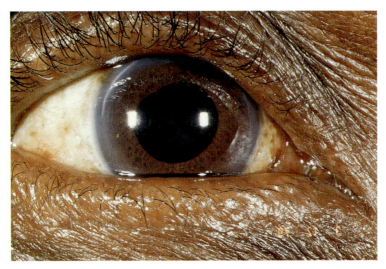

FIGURE 4.5   An example of a clear opening left after most of a traumatic cataract has dissolved and a gray membrane remains. (The clear opening is to the left of the corneal reflection, which is on the right.) (Courtesy of Robert Stegann, M.D.)

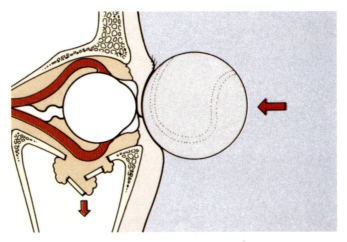

FIGURE 4.8   Drawing of a blowout fracture. In this case, a blunt injury to the eye has produced a fracture in the floor of the orbit.

FIGURE 4.9 This patient shows the accumulation of blood in the lids that comes from the torn vascular structures in the orbit.

secretions. Therefore, preoperative use of anti-inflammatory drugs, plus very delicate manipulation of eye tissue during surgery, inhibit further natural responses and make the outcome of the surgery more predictable.

If the young man's eye had not been surgically repaired, nature's own repair would probably have afforded him some sight in that eye. However, modern surgical repair allowed him to read the 20/40 line with the proper spectacles.[4]

3. CORNEAL LACERATION AND TRAUMATIC CATARACT IN A CHILD

During one of my trips to Africa, a little boy was struck in the eye with a sharpened pencil and rushed to the clinic. Because the child was crying and in pain, we could not examine his eye in the clinic. In order to perform the examination carefully, the child was taken to the operating room and placed under general anesthesia. First we noted the corneal laceration with gray swollen edges.* Here too, the iris had prolapsed into the wound where it was held securely. The pencil tip had obviously gone deeply into the eye, penetrating the lens and forming a cataract. Happily, the eye could be repaired surgically. However, from examining many such children who did not come to the clinic for immediate treatment, we now know how the eye would have healed naturally. In the natural situation, the corneal and iris response seals the wound as noted in the last case. Now, let us look closely at the lens reaction. Once the capsule of the lens is pierced, the lens proteins agglutinate, and the previously clear lens becomes a cloudy gray. Slowly, over the course of weeks, the lens protein sets off an acute inflammatory reaction, as if the body was challenged by a foreign protein. Admittedly, reacting to one's own tissue in such an aggressive way is unusual. From experiments, we have learned that the lens proteins are considered to be a foreign substance by the body. But why should the native protein in one's own lens be seen as foreign? It's time for another speculative hypothesis. In the developing embryo the blood circulation to the developing lens closes down at the 20th week of gestation, that is, the blood vessels that originally surround and nourish the embryologic lens close down and then disappear. Thus, the lens material is never exposed to the circulation and never gets registered as native body protein during the crucial late fetal period when native structures are registered by the immune system as "self."[4] If the lens capsule is pierced during injury, this "foreign" lens protein is exposed to the body. The body attacks by sending in special white blood cells. These cells systematically envelop the lens protein, remove it, and lay down a type of scar tissue (i.e., a traumatic cataract or an opaque lens capsule). Interestingly, the opening made by the wounding missile (which pierced the anterior and posterior lens capsule) may remain within the midst of the traumatic cataract (Figure 4.5 in

* When the cornea swells significantly, it loses its clarity, scatters more light, and becomes gray in color.

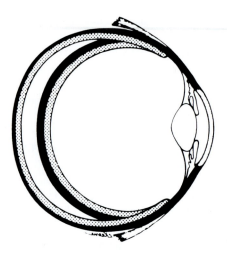

FIGURE 4.6   Once the eye lens has been injured, it no longer acts as a focusing element. To compensate for the absence of the lens, the eye gets longer. (From Miller D: *Optics and Refraction.* Gower Publishing, New York, 1991, p. 86, with permission.)

color insert). Although a clear opening will allow light to reach the retina, absence of the normal lens means that the light will not be focused sharply onto the retina. How does the body cope with this problem? Studies show that in a child under 10, the eye with a traumatic cataract will slowly elongate over the next year or two.[5] This elongation allows the out-of-focus rays of light to become better focused on the retina (Figure 4.6). Thus, in the optimal healing scenario, the child's eyeball elongates so that the cornea alone will be able to focus the light onto the retina through the small clear hole (produced by the missile) within the cataract. The presence of a small hole in the opaque capsule helps by producing an increased depth of focus.

These body responses place in new perspective the breadth of evolution's role in individual survival. Of course, this response was modified in the patient who was surgically repaired. The traumatic cataract was extracted, which prevented the autoimmune response. The corneal laceration was repaired and an intraocular lens was placed in the position of the original lens, so that normal focusing could take place. Under these conditions, the eye will probably not elongate.

4. CATARACT AND IRIDOTOMY
FROM BLUNT TRAUMA

During another trip to South Africa, I was also able to follow the unusual case of a young woman struck in both eyes by a bullwhip. Although we could never

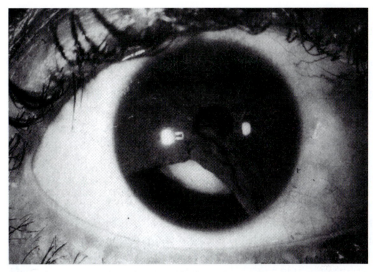

FIGURE 4.7   The eye of the patient that suffered a bullwhip injury. Note the white cataracts and the triangular opening at the iris root, which will function as the new pupil. (Courtesy of Robert Stegmann, M.D.)

confirm the information, the rumor was that she had been beaten by her jealous lover. Fortunately, the corneas were not pierced. However, the blunt force of the whip produced cataracts in both eyes. The blow also tore the inferior root of the iris in each eye, allowing light to reach the retina through a peripheral iridotomy beyond the edge of the cataract (Figure 4.7). Without the power of the lens, the light going through the new opening was not sharply focused on the retina. Nevertheless, she ultimately was able to see clearly out of both eyes with the proper aphakic spectacle correction without the need for surgery. Her clear vision with the proper glasses also tells us that the edge of the cornea (an area not normally used to focus light) has useful optical properties.[6] The combination of the peripheral cornea, the spectacle lenses, and probably some visual brain enhancement yielded the final 20/60 visual result.

Let us examine the events that took place as the whip struck the eyes. The blow momentarily indented the eyes and created a high pressure wave that produced a rip at the root of the iris. The contraction of the sphincter muscle surrounding the pupil then pulled away the torn iris from its root, creating triangular openings at the edge of iris (the iridotomies).

The cataracts were also produced by the powerful blow to the eye. Such a blow sends a shock wave through the eye that induces wrents in the membranes of the individual lens fibers. This injury disrupts the ability of the lens to keep the high protein concentration in clear solution. The proteins aggregate, and increased light scattering is the result. Although the lens capsule remains intact, a cataract develops. On further review of the iris anatomy, something unexpected was noted. In the human, the iris root is thinner and weaker than the rest of the iris. This finding suggests that the iris root will tear first when a blunt force produced by a whip, a fist, or a thrown rock strikes the eye. Of course, such a blow also produces a cataract. Thus, the eye responds to a blow of this magnitude by developing both a cataract and the compensatory clear opening at the edge of the cataract. This chain of events might be considered nature's version of "cutting one's losses". Interestingly, the thin iris root is not seen in most mammals, but is found in primates and humans.

5. BLOWOUT ORBITAL FRACTURE

This story involves a nurse who worked in our clinic in Boston. One fateful evening, she was involved in a head-on auto collision. Upon impact, her face struck the dashboard. Superficially, she seemed to come out of the accident with good vision, only mild pain, a black and blue eye, and only minor cuts and bruises to the face. However, when we examined her X-rays, we saw that the floor of the orbit (blowout fracture) had been fractured, and there was blood in the maxillary sinus. What happened?

From anatomic studies, we know that parts of the orbital floor are paper thin. Actually, it almost looks as if the floor was designed to give way in case of blunt injury. Studies show that as a blunt object strikes the eye, it forces the eye deep into the orbit and also presses the eye into an elongated, vertical elliptical shape. As the elongated eye is rammed against the orbital floor, the thin floor absorbs much of the force of the blow and cracks (Figure 4.8 in color insert). This chain of events suggests a design in which the orbital floor is sacrificed in order to preserve the integrity of the eye.

A blow to the eye and orbit often tears blood vessels behind the eye and produces an intraorbital hemorrhage. The blood trapped behind the eye could compress the optic nerve, strangle the fine nerve fibers, and produce blindness. Fortunately, this rarely occurs because of the unusual design of the orbit.

The blood vessels in the orbit are organized into a system that readily diverts free orbital blood out to the lids and/or through the cracked orbital floor into the underlying maxillary sinuses. Absence of valves in the veins, as well as the strategic location of planes of connective tissue in the orbit, act as conduits that lead the blood away from the eye and optic nerve toward the lids as well as through cracks in the ceiling of the maxillary sinus.[7] That is why the patient displayed swollen, black and blue lids, as well as a collection of blood in the maxillary sinus (Figure 4.9 in color insert).

Hence, although sustaining a severe accident, the orbital structure prevented loss of vision. The accident produced no facial or eye deformity and no entrapment of the eye muscles that would have required surgery. She was back to work in a week.

6. BLUNT TRAUMA AND THE
OCULO-CARDIAC REFLEX

The patient to be described demonstrates an unusual body response when the eye is injured. During the repair of a child's strabismus in the operating room the anesthetist looked up at me and announced that the child's pulse had suddenly slowed down. Every time I grabbed the medial rectus muscle to move it to the proper location, the pulse slowed in the same fashion. Happily, once I had sutured the muscle onto the sclera, the pulse returned to normal levels. A few weeks later, I saw the same slowing of the pulse occur when an eyelid injury was repaired. The reflex is well known and is called the oculo-cardiac response.[8–10] What's going on?

One of the major cranial nerves, (the vagus nerve) sends branches to the muscles of the eye, as well as the eyelid. This same nerve also sends branches to the heart in order to control heart rate. When touched, pressed, or pulled, these branches to the ocular area can slow the heart rate by as much as 50%. Therefore, the oculo-cardiac reflex can be seen during eye muscle surgery, eyelid surgery, or can sometimes occur if you simply press on the eye.[8–10] What benefit to survival could such a reflex have?

A slow pulse rate means a diminished blood flow during a hemorrhage. A slower blood flow can allow a hemorrhage to be stopped more easily by natural clotting. By slowing the flow of blood, the elements that must mix to produce a clot have a better opportunity to combine. The pulse-slowing oculo-cardiac reflex suggests that the eye area is favored during injury. Typically, a hemorrhage in the area of the eye during surgery or from an accidental injury seems to clot faster than from other parts of the body. Therefore, a hemorrhage in the eye area will stop sooner and probably prevent more serious injury to structures such as the optic nerve.

7. THE OTHER EYE (THE ULTIMATE
PROTECTION AGAINST BLINDNESS)

Thus far we have seen how the body can restore good visual function if an ocular injury is superficial, and useful function if the injury is more severe. How is vision restored if the injury to the eye is catastrophic? Happily, we can always turn to the other eye.

How easy is it to turn to the use of the remaining eye after one loses an eye? I'd like to quote from a letter I received from J. Stuart Freedman, Jr., Editor of the journal *Ophthalmology.*

> In September of 1989 I had my left eye taken out at the Mass Eye and Ear Infirmary. Two days post-op I left the hospital to return to Philadelphia. As my wife and I went to South Station I was in a state of near panic—my field of vision was so restricted that everything was as seen through an upended mail slot. Anger warred with claustrophobia. But in less than two months the right eye apparently stopped trying to find the left eye. Incredible and with great relief, it has remained so ever since. I have no sense of my field of vision ever having been broader than it is now.

There have been other prominent people who have gone on to continue their lives with great success after losing an eye. We know that Alexander the Great lost an eye, as did Lord Nelson, and the Israeli general Moshe Dyan. Teddy Roosevelt, while a student at Harvard, was struck in the eye while boxing and suffered a detached retina and ultimate blindness in that eye. James Thurber, the humorist, lost one eye to an arrow injury when young.

Recent studies in lower vertebrates (frogs and chicks)[11] teach us that there is at least one gene, and possibly a second, which orders a single band of tissue in the embryo (potential eye tissue) to divide into two and go on to produce the two eyes. Thus in the same genetic plan to have a visual organ is the plan to provide for a second organ.

SUMMARY

A. NATURAL EYE REPAIR AND THE COMMUNITY

By presenting these examples of eye injury, I have tried to give the reader some insight into the sophistication and majesty of the eye's own repair system. Admittedly, the repair system is not unique to humans. It is a system that has developed during evolution. Of course, it does not produce the repair as perfectly as that performed in the modern operating room. However, in the most severe of injuries, it is capable of producing "getting around" vision. The high level of performance of the natural system seems to explain why the high incidence of eye injury in human society does not lead to an equally high incidence of blindness.

The term "getting around" vision was used in the last paragraph. Let us explore the implications of that term. Most of the time, the body's repair of a serious eye injury does not lead to perfect vision (i.e., 20/20). Such severely injured eyes, if left to heal on their own, may achieve a visual acuity only about one-tenth the 20/20 level. If this injured eye is the better eye, then the patient would be legally blind by Western standards. Getting around vision is equivalent to less than 20/200 vision. Patients with getting around vision can function in their local setting, performing all their duties as parent, cook, and provider. The term local setting implies that they are familiar with the local landmarks, people, and customs of their community. I do not think these victims could survive if left on their own in the wild for extended periods. Hence, the eye's repair system for serious injuries must work in conjunction with a friendly community to permit survival of the victim.

B. EYE HEALING IN CERTAIN ANIMALS

One can only roughly estimate the level of functional success of the natural eye repair system when it is combined with the resources of the human community. On the other hand, let us look at the eye repair system of the newt. It is able to produce a brand new lens if its original lens is damaged. The rabbit can reproduce a brand new layer of corneal endothelial cells and control corneal transparency when its cornea is damaged. If the optic nerve of the fish is severed, it will regrow. The human system can do none of the above. Consequently, the ultimate level of natural healing of serious ocular wounds in the human is rarely perfect. However, I am suggesting that we have not had to evolve a system of healing like the newt, rabbit, or fish, because of the support of our societal structure, as well as the high level of human intelligence.

REFERENCES

1. National Society to Prevent Blindness: Fact Sheet, New York, 1980.
2. Shingleton BJ, Hersh PS, Kenyon KR: *Eye Trauma.* Mosby-Year Book, St. Louis, 1991.
3. Tabbara KF: Prevention of blindness in developing countries. In *Prevention of Eye Disease* (Friedlander M, ed.). Mary Ann Liebert Publishing New York, 1988.
4. Miller D, Stegmann R: *Treatment of Anterior Segment Trauma.* Medicopea, Montreal, 1985.
5. Rasooly R, Ben Erza D: Congenital and traumatic cataract: The effect on ocular axial length. *Arch. Ophthalmol.* 106:1066, 1988.
6. Miller D, Atelbara N, Stegmann R: The role of the limbal cornea in vision. *Eye* 3:123, 1989.
7. Koomneef L: Spatial aspects of the orbital vascular system. In *The Biomedical Foundations of Ophthamology,* Vol. 1 (Tasman W, Jaeger EA, eds.). Lippincott, Philadelphia, 1990, ch. 33.
8. Aschner B: Uber ein bisher noch nicht beschriebenen reflex von Aug Suf Krieslauf and Atmung. Verschwinden des Radialispulses bei druck suf das auge. *Wein. Klin. Ochenschr.* 44:1529–1530, 1908.
9. Gunson EB: The oculo-cardiac reflex. *Br. J. Child. Dis.* 12:97–105, 1915.
10. Levin SA: The oculo-cardiac reflex. *Arch. Intern. Med.* 15:648, 1915.
11. Bower B: New gene clearly resolves an eye debate. *Science News* 151:102, Feb 15, 1997.

5

REFRACTIVE ERRORS OF THE HUMAN EYE: A SOCIOLOGIC VIEWPOINT

Introduction
1. Myopia
   A. *Optical Considerations*
   B. *Related Factors*
   C. *Cultural Considerations*
2. Astigmatism
   A. *Optical Considerations*
   B. *Related Factors*
   C. *Cultural Considerations*
3. Hyperopia
   A. *Optical Considerations*
   B. *Related Factors*
   C. *Cultural Considerations*
4. Presbyopia
   A. *Optical Considerations*
   B. *Related Factors*
   C. *Cultural Considerations*
Summary
   A. *Refractive Errors and Society*

INTRODUCTION

In an earlier chapter, I discussed the creation of the retinal image produced by the optics of a typical healthy and normal human eye. Typical is a difficult concept, because about half the American population wears glasses. Is the typical eye emmetropic, myopic, astigmatic, hyperopic, or presbyopic?

This chapter will not only look at all these refractive conditions, but ask the question, Could these refractive conditions prove useful to the survival of a social group, particularly an ancient or primitive group?

1. MYOPIA

A. OPTICAL CONSIDERATIONS

In a myopic eye, the optical power is too strong or the length of the eyeball is too long.[1] More specifically, the power of the cornea and lens is not coordinated with the eyeball length. In Figure 5.1 we can see that the parallel rays entering the eye represent rays from some distant object. The myopic eye focuses a sharp image in the vitreous cavity in front of the retina, and so an out-of-focus image is registered on the retina.* On the other hand, near objects are in focus on the retina.

B. RELATED FACTORS

The refractive surgery business is projected to reach into the billions of dollars by the year 2000. Thus it comes as no surprise that investment companies have analyzed the potential North American myopic market. Roughly 70 million people or almost 25% of the population fall into that category. On the other hand, the incidence of a comparable degree of myopia in a group of representative Melanesians from the New Hebrides islands was less than 5%. However, studies show that close to 85% of Taiwanese are myopic.† These findings suggest that genetic factors influence the development of myopia.[2-4] Racial background as well as one's own family history influence the incidence of myopia. A recent study[5] noted that the children of myopic parents had a statistically longer eye (axial length measurement) than the children of a control group.

However, another important factor that is closely associated with myopia is the amount of near work done in a young person's formative years. This idea got started over 100 years ago. In 1866, German ophthalmologist Herman Cohn studied 10,000 children living in the city of Breslau.[6] He concluded that myopia increased in both prevalence and amount as children progressed through school. Recent studies have confirmed Dr. Cohn's conclusions. For example, studies in

---

* Higher degrees of myopia require stronger minus lenses. Since a strong lens has a shorter focal length, the French ophthalmologist Monoyer thought it more logical to give a stronger lens a higher value. To do this, he suggested taking the reciprocal of the focal length. This was approved at the Heidelberg Ophthalmological Congress in 1875.

† The incidence of myopia in Taiwan (as elsewhere) varies with age. Thus, myopia is present in about 12% of those 6 or younger; 55% in those 12 or younger; 76% in those 15 or younger, and 84% in those over 18. (Luke L-K lin, et al: Epidemiological study of ocular refraction among school children in Taiwan. Invest. Ophthalmol. Vis. Sci. February 15, 1996, 6(3):51002.)

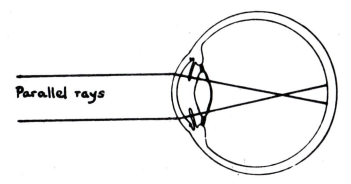

Parallel rays

FIGURE 5.1   In the myopic eye, parallel rays of light coming from a distant object come to a focus in front of the retina.

the United States show a strong relationship between prevalence and amount of myopia with years of graduate education. Informally, I confirm these findings every year when I survey second- and third-year medical students and consistently find a 60% to 70% incidence of myopia. All of the above information suggests that a formula that includes cumulative time spent in close work and a genetic proclivity will determine who will become myopic and by how much.

There is, however, a small subset of the myopic population with very high amounts of nearsightedness (over 6 diopters or a far point of 16.7 cm or closer) who do not fall into this "increased near work" group. Almost all of these "high myopes" have significantly elongated eyeballs. Clearly this type of myopia has a much more powerful genetic influence.

C. CULTURAL CONSIDERATIONS

The information described so far suggests that in early human societies, there would have been a small group of myopic people. Most of them would have naturally drifted into near tasks, such as weaving and net making. As this group aged and became presbyopic, their nearsighted status allowed them to continue to see clearly their near tasks. A second subset of myopes would have been the high myopes. Their number would be quite small. Not being able to see clearly beyond a few inches, they would be very dependent on the other tribal members. However, by being able to see objects very close to them, they possessed an innate magnifying system (Figure 5.2).

Upon a visit to National University of Taiwan, my host took me to the National Museum in Taipai. One of the most striking exhibits was that of the Netsuke statues. These are statues about one inch high and show surprising detail. Figure 5.3 (see color insert) is an example of such a statue. A large magnifying glass hovers over the exhibit so that the visitor can appreciate the quality of the figures. Obviously, some form of magnification was needed to create these sculpted figures.

There are 3 ways to make the retinal image larger
1. Use a larger object

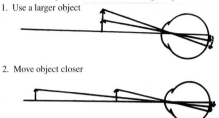

2. Move object closer

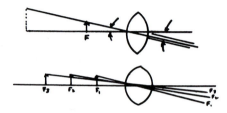

3. We can enlarge the object with a lens
   also put the image at infinity so accommodation will be no problem

FIGURE 5.2   A diagram showing that the closer an object can be held to the eye, the more magnified the retinal image.

Yet there is no evidence that the ancient adult artists who created these statues had access to magnifying devices. A logical explanation would be that highly myopic (nearsighted) artists (or very young artists with a large accommodation range), could have held the pieces of ivory close enough to their eyes to have been able to carve the tiny details onto such small statues.

   People with high degrees of myopia have very large eyes. You may recall that it was noted that the eagle achieves a magnified image of distance objects because of its large eye. As opposed to the eagle, the high myope cannot see clearly in the distance without correcting spectacles, but is able to see magnified versions of objects held very close for two reasons. First, the closer an object is to any eye, the larger the retinal image. Second, the retinal image is larger in the larger myopic eye. Unhappily, there is a dangerous down side to high myopia. Patients with very high degrees of axial myopia also stand a greater chance of developing a retinal detachment. Prior to modern treatment, a retinal detachment often led to blindness in that eye. This may explain an old Chinese proverb that claimed that those who performed the forbidden stitch in embroidery would go blind. The forbidden stitch was an extremely fine stitch, which could only be performed by someone who had the innate magnification of a highly nearsighted person, and of course, such people were more prone to retinal detachments and subsequent blindness.

   On a less serious note, I will never forget the scholarly myopic librarian at the medical library of the Massachusetts Eye and Ear Infirmary who would periodically

FIGURE 5.3   Note the detail carved into this Nietzke statue, which only stands 1 inch high.

FIGURE 5.4   The shape of a spherical cornea as opposed to an astigmatic cornea can be compared to a round spoon vs. a tablespoon.

FIGURE 5.5 This picture demonstrates the way a colored ball would look to **(a).** a normal subject, **(b).** a subject with vertically oriented astigmatism (with-the-rule astigmatism), and **(c).** a subject with oblique astigmatism. (From Mueller CG, Rudolph M, The Editors of Time-Life Books: *Light and Vision.* Time Life Books, New York, 1966, p. 86, with permission.)

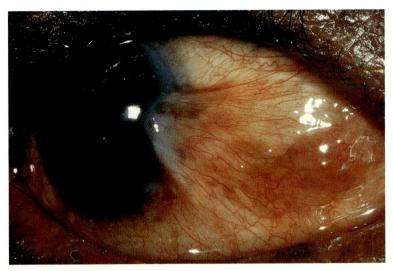

FIGURE 5.7   This eye has a fleshy lesion that goes from the conjunctiva to the cornea. The lesion is known as a pterygium. It can induce up to 3 diopters of astigmatism (with-the-rule astigmatism).

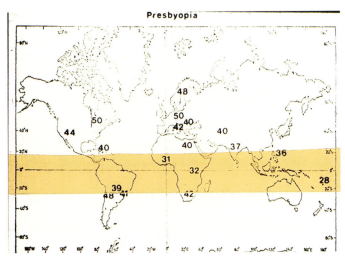

FIGURE 5.11   This world map shows that the age of onset of presbyopia increases the farther one gets from the equator. (Modified from Weale R: *The Aging Eye: Geographic and Climate Influences on the Aging Process.* Second International Symposium on Presbyopia, Essilor Ltd., Paris, 1981, with permission.)

take off his thick spectacles. When I asked him why he did that, he replied, "When the pressure gets too great, I like to simply reduce the outside world to one big blur."

2. ASTIGMATISM

A. OPTICAL CONSIDERATIONS

Astigmatism is a type of refractive error in which the shape of the cornea distorts the focusing ability of the eye.* Instead of being spherical and focusing rays from all directions to one focal point, the astigmatic cornea might refract light rays in a vertical plane differently than those, for example, in a horizontal or oblique plane. An astigmatic cornea would be shaped more like an elongated tablespoon than a rounded tablespoon (Figure 5.4 in color insert). In Figure 5.5 (see color insert) we can see how a ball looks to someone with no astigmatism (round and sharply outlined), to someone who has a vertical type of astigmatism (a blurry, vertically elongated ball with accentuated vertical lines), and finally to someone with an oblique type of astigmatism (tilted, blurry ball).

B. RELATED FACTORS

As noted in the chapter on infant vision, almost 60% of infants from birth to 2 years of age have been measured to have significant amounts of astigmatism. In Figure 5.6 we can see how the amount of average astigmatism plummets downward as the infant ages.[4] If we define a significant amount of astigmatism to be 2 diopters or more, then only about 10% of the adult Western population would fall into that category. If 3 diopters or more were the cutoff, then only about 3% of the same population would fit into that group.[2,7]

Most astigmats inherit their condition. However, there are people who develop significant amounts of astigmatism after penetrating eye injuries or after eye surgery. Finally, there is another geographic group with a high incidence of astigmatism. These people have a condition known as pterygium (Figure 5.7 in color insert). Pterygium is common among people living in warm climates,[8] actually affecting about a third of the population living in the tropics.† From a mechanical point of view, the pterygium can be thought to anchor itself on the conjunctiva

---

* Another ocular element that may contribute to astigmatism is a tilted or asymmetric position of the crystalline lens.

† Why are pterygia so common in the tropics? While working in Samoa, I noted that the pterygium seemed to act as a wedge-shaped plug that prevents perspiration from getting into the eye. It would appear that the irritating factors in sweat initially get into the eye and stimulate pterygium growth. The pterygium could be a response that prevents further runoff of perspiration into the eye, particularly if the eyes are squinting in the bright light. You might ask from a survival point of view, is the benefit of keeping sweat out of your eyes while you are squinting worth the astigmatic distortion produced by the pterygium. Happily, the act of squinting (narrowing the palpebral aperture) is equivalent to inducing a horizontal stenopaic slit, which optically corrects the "with-the-rule astigmatism."

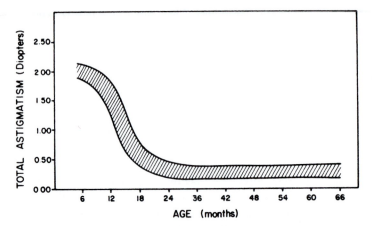

FIGURE 5.6 A graph comparing the average amount of astigmatism with age. Note that the average astigmatism for an infant is 2 diopters. By age 2 this amount has plummeted to about 0.5 diopters. (From Grosvenor T, Flom MC: *Refractive Anomalies. Research and Clinical Applications.* Butterworth-Heinemann, Boston, 1991, with permission.)

and pull on the corneal curvature in a horizontal direction, thus creating a steeper corneal curvature vertically. Many of the patients with pterygia of moderate to large size averaged about 3 diopters of "with-the-rule astigmatism." Their specific type of astigmatism would tend to distort the world by elongating and accentuating vertical objects.

C. CULTURAL CONSIDERATIONS

While lecturing to a group of older ophthalmologists a few years ago, I was asked a strange question. One of the doctors in the back of the room stood up and told the audience that during World War II, he was a sailor on a convoy ship. He recalled that the captain of the ship always chose sailors who had astigmatism to be lookouts. He wanted to know why astigmatics rather than young men with normal vision would be recruited to be the lookouts.

During World War II, the Nazi submarine was the great predator of the Allied convoy ships. At that time, prior to the use of sonar, the only way to know that a submarine was in the area was to see its periscope. An able lookout would have to see the dark vertical tube of the periscope against the background of the gray sea.

With-the-rule astigmatism creates a distorted retinal image (similar to that created by the pterygium) in which vertical objects such as a periscope are highly accentuated (Figure 5.5).* Therefore, people with that form of astigmatism would

---

* In fact, the eye practitioner takes advantage of this accentuation when he or she presents a 360° target of radiating lines to the astigmatic patient. The direction of the radiating lines that look the blackest determine the major meridian (not the axis) of the astigmatism.

be optically suited to detect the presence of the periscope of an enemy submarine better than normal sighted people.

If a primitive society had a few such astigmats, they could well serve as lookouts, and probably see the vertical forms of attacking enemy soldiers before anyone else in the tribe.

3. HYPEROPIA

A. OPTICAL CONSIDERATIONS

Most children are hyperopic, or farsighted. Optically speaking, light rays from a distant object are brought to a focus behind the retina of the hyperopic eye (Figure 5.8). Simply stated, this occurs because the focusing structures (cornea and lens) are too weak or the axial length of the eye is too short. The result is an out-of-focus image on the retina. However, as opposed to the myope, the hyperope can refocus the blurred image by increasing the optical power of the lens of the eye (accommodation). The hyperope simply takes advantage of this same process to refocus a distant object onto the retina. Therefore, the average amount of hyperopia is a compensable condition in a young person with a sizable range of accommodation. As opposed to the emmetrope, the hyperope must almost "waste" accommodation to get distant objects in sharp focus, and then use additional accommodation to see a close object clearly. As time goes by, hyperopes tend to need reading glasses at an earlier age than their emmetropic colleagues. In primitive societies in which there were no reading glasses available, the middle-aged hyperopic artisan would lose the ability to follow the fine details of his or her weaving sooner and probably depend more on touch and feel than the other weavers.

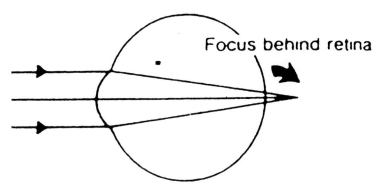

Hyperopia (farsightedness)

FIGURE 5.8   In the hyperopic eye with the accommodation relaxed, parallel rays from a distant object focus behind the retina.

B. RELATED FACTORS

About 50 million North American adults are farsighted.[4] However, since most young hyperopes can compensate for this condition by using accommodation, only a small number of hyperopes require spectacles or contact lenses.

C. CULTURAL CONSIDERATIONS

Farsighted people are more vulnerable to another eye condition. Although the incidence of esotropia in our culture is about 1%, most people with esotropia are hyperopic. In older cultures, people with crossed eyes were often felt to be possessed by demons. In many such societies, people with physical deformities, such as crossed eyes, are treated differently by other tribal members. They are usually treated as social outcasts and so may develop disturbed identities. Such people develop their own sense of reality and often get involved with the spiritual world. In time some may become the shaman or medicine man of the tribe.

4. PRESBYOPIA

Presbyopia or "old age sight" describes the inability to accommodate or refocus in order to see close at hand objects such as newsprint, a fine crochet pattern, fishing nets, and so on.

A. OPTICAL CONSIDERATIONS

As we age, the accommodative system of the eye changes in three fundamental ways. First, the lens of the eye continues to add layers of tissue much like a tree adding rings. It is thought that since the lens cannot discard old, damaged tissue layers into the eye fluids much as the skin sloughs dead layers into the air or the snake molts, aged lens tissue is transported inward toward the lens center, while fresh outer rings are produced. The result is that the lens of the eye continues to enlarge, actually tripling its volume during the journey from infancy to old age. The larger and denser the lens becomes, the less elastic it becomes. The larger size, lowered elasticity, and greater proximity to the ciliary processes prevent easy change of shape (accommodation) when acted upon by the ciliary muscle (Figure 5.9). Figure 5.10 is a simple cartoon that schematically suggests how the ciliary muscle affects the change of lens shape.

Second, as the normal sphincterlike ciliary muscle contracts, it draws closer to the edge of the lens. In this process, some of the fibers or zonules that stretch from the ciliary body to the lens slacken, and their pull on the lens diminishes, while those fibers closest to the lens equator may actually tighten.* With this new distri-

---

* The idea that equatorial zonules might actually tighten during accommodation was suggested in Schachar R, et al: The mechanisms of accommodation and presbyopia in the primate. *Ann. Ophthalmol.* 27(2):58–67, 1995.

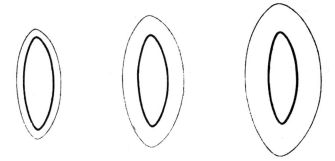

FIGURE 5.9   Anatomy of aging. As the lens ages, it gets larger and thicker, thus crowding the ciliary processes and making the anterior chamber shallower.

bution of tension, the center of the lens tends to assume a more natural, rounder shape. A rounder, more curved lens center creates additional focusing power of accommodation. As the eye ages, the ciliary muscle becomes less efficient. The result is a further reduction of accommodation. To add the final blow, the fibers or zonules change their position on the lens as it grows. In so doing, the angle of pull by the muscle is less effective.

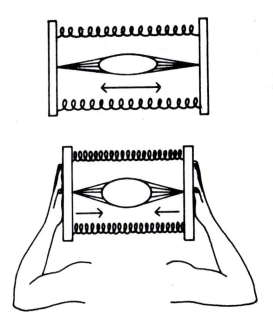

FIGURE 5.10   This diagram demonstrates the effect of contraction of the ciliary muscle on accommodation. As the tension on the zonules slackens, the lens goes to a more curved shape, increasing its focusing power. (From Miller D: *Ophthalmologica.* Editorial Limusa, Mexico, 1983, p. 20, with permission.)

This process continues to occur throughout life. As children, we start with approximately 15 to 20 diopters of accommodation. This means that a normal eyed child can focus on a tiny object about 2 $^1/_2$ inches ($^1/_{15}$ meter) from its eyes. By the age of 45, we have lost all but about 3 diopters, and by the age of 60, almost no accommodation is left.

B. RELATED FACTORS

A table from a publication by the U.S. Department of Health, Education and Welfare describes the percent of the population wearing glasses and contact lenses.[9] In the age group of 45 to 54, the number jumps to about 80%. In a most dramatic manner, this table tells us that presbyopia and the need for reading glasses starts after age 45. This number applies to the average North American. Interestingly, the age of onset of presbyopia varies over the world. In warmer climates, people become presbyopic in their mid-30s (Figure 5.11 in color insert). In cold climates, the onset of presbyopia occurs in the late 40s. Therefore, one might speculate that in ancient African communities, the onset of presbyopia would have also occurred in the mid-30s. It is interesting to speculate that in ancient times the problem of presbyopia was far less important, since most humans did not live past the age of 40.

C. CULTURAL CONSIDERATIONS

While I served as a volunteer ophthalmologist in Colombia, I would often travel into the rural areas to examine patients. In such outlying areas, far from the city, middle-aged patients used children to do their reading when they became presbyopic. Only those who had access to an eye clinic or a spectacle vendor would be able to read themselves. What struck me was the implication of this use of children in the ancient village, before the era of reading glasses. Youngsters must have been recruited to stay with the elders and help them with their near tasks, (e.g., weaving, utensil making). In a sense, this implied an enforced association between the young and old of the village. This association could have served as an opportunity to transmit the community's culture from generation to generation.

To many people today, the advent of presbyopia and the need for reading glasses signals the start of old age, a time fraught with negative connotations. It might soften the blow if we would look at such a time as the ancients did, that is, a time when the youth of the village would come to the elders in order to guarantee that tribal knowledge would continue to survive. In fairness, it could be argued that the infirmity of arthritis of the elderly also forced the children of the village to spend more time with them. Thus, in a human social setting, certain chronic conditions of the elderly were redirected to have positive influence on the cohesion of the group. Oliver Sacks has aptly called this situation "the creative potential of disease.[10]

SUMMARY

A. REFRACTIVE ERRORS AND SOCIETY

In any given ancient community (pre-spectacles), most of the people would probably have been normal sighted (emmetropes) or mild hyperopes. However, sprinkled in among the normal population would be some myopes. They would be less apt to be the warriors, but would be better suited for carving, toolmaking, and weaving. Since myopes are always in focus for near tasks, they are less bothered by presbyopia and can continue their near work throughout their lives.

On the other hand, the emmetropes and hyperopes of the village would need the help of the children to perform certain near tasks when they became presbyopic (if they lived to the age of presbyopia).

It was further suggested that presbyopia might be exploited by a well-knit social group in bringing the old and the young of the village together in order to help transfer the culture of the village from one generation to the next.

Finally, it was suggested that the occasional person who had a significant amount of "with-the-rule astigmatism" saw vertical figures in an exaggerated manner. That person could well have been the ideal lookout, since seeing the vertical forms of hostile enemies at a far distance would have a distinct survival advantage for a tribe.

If all of the arguments have any merit, then one could speculate that the natural distribution of refractive errors in the primitive communities helped foster a division of labor that gave the community additional tools to enhance their survival under evolutionary stress. Then one might think of the distribution of refractive errors in a human community as yet another cord that binds our social interdependence.

REFERENCES

1. Stenstrom S: Untersuchange uben die Variation and Kovariation des optisches. Elements des menschlichen Auges. *Acta Ophthalmol. Suppl.* 26 (also English translation by Woolf D: *Am. J. Optom.* 25:218–232, 1948).
2. Bennett AG, Rabbets RB: *Clinical Visual Optics,* 2nd Ed. Butterworths, London, 1989.
3. Gile, GH: The distribution of visual defects. *Br. J. Phys. Opt.* 7:179–208, 216, 1950.
4. Grosvenor T, Flom MC: *Refractive Anomalies Research and Clinical Applications.* Butterworth-Heinemann, Boston, London, 1991.
5. Zadnik K, Satariano WA, Mutti DO, et al: The effect of parental history of myopia on children's eye size. *JAMA* 271:1323, 1994.
6. Cohn H: *Unter der Augen von 10,000 Schulkindern nebst Vorsehlangen zor Verbesserung der Augen Nachteilligin Schuleinrich tungenz.* Eine Atiologische Studie, Leipzig, 1866.
7. Waring G: *Refractive Keratotomy for Myopia and Astigmatism.* Mosby Year Book, St. Louis, 1992, pp. 1075–1076.
8. Cameron ME: *Pterygium Through the World.* Charles C Thomas, Springfield, IL, 1965.
9. National Health Survey. Refraction Status and Motility Defects of Persons 4-74 years. United States, 1932–1972, Hyattsville, MD, U.S. Department of Health, Education and Welfare, HEW pub. no. (PH) 78-1654, 1978.
10. Sacks O: *An Anthropologist on Mars.* A.A. Knopf, New York, 1995.

6

EYE COMMUNICATION

Introduction
1. The Human Eye
   A. *The Pupillary Response*
   B. *Crying*
   C. *Eye Reddening*
   D. *Eye to Eye Contact*
   E. *Blink Rate*
   F. *Positioning of Brows and Lids*
   G. *Determining Age Through the Eyes*
   Summary
   A. *Advantages of Eye Communication*
   B. *Eye Communication in Painting*
   C. *Determining Health Through the Eyes*
2. Animal Eye Spots
   A. *Butterfly Eye Spots*
   B. *Eye Spots in Other Species*
   Summary
   A. *Thoughts on Animal Eye Patterns*
   B. *Thoughts on Manmade Eye Patterns*

INTRODUCTION

In an earlier chapter, we learned that although the infant does not possess acute vision, it is quite expert at recognizing different facial expressions. Studies[1] have also shown that when pictures of different line drawings of faces are presented to an infant, only faces that are round and have two symmetrical eyes are appreciated as faces. As the infant develops into an adult, the presence of the eyes not only helps define a face, but they are also used to convey different levels of emotion and subconscious thoughts to the observer.

Therefore, in this chapter, we will look at the way in which the eyes, the brow, and the lids are used as transmitters of information and emotion. Clearly, this ability to communicate nonverbally was important to the survival of our human ancestors.

1. THE HUMAN EYE

A. THE PUPILLARY RESPONSE

I was introduced to the area of emotional pupillary response in an odd way. An executive from a large New York advertising agency called me unexpectedly. It seems that the company wanted to develop a method of objectively evaluating responses of prospective customers to different T.V. commercials. Up until that time, their practice had been to show samples of their commercials to volunteer audiences, composed of typical homemakers. Unhappily, the agency had discovered that these subjects did not want to offend the agency by criticizing any of their commercials. Thus, the agency could not get the frank evaluation that they sought.

One of the company executives had been reading the psychology literature and had discovered that pupillary dilation occurred almost instantly after a surge of adrenaline was released into the bloodstream. This meant that registering pupillary enlargement was a simple method of tracking blood adrenaline and might represent an objective sign that the individual was excited. His review also uncovered the fact that a colleague and I had built a small pupillometer.[2] Consequently, an engineer and I were hired to build a nonintrusive device that could follow the pupil movements of their selected volunteers. As we started our work we learned that a major dilemma that the advertising industry faced was that the average American is bombarded by hundreds of newspaper, billboard, and T.V. commercials each day. Therefore, most people become immune to their effect. Only something really new and startling will grab attention. When that happens, the pupils enlarge.

Of course, we did not invent psychological pupillometry. It had been used by American intelligence officers when interrogating spies during the Vietnam war. Pupil changes are like the changes picked up in a lie detector test. When the average person deliberately lies, they become internally uncomfortable and a surge of adrenaline is released. This causes an increase in pulse rate, blood pressure, sweating, and pupil dilation. In the typical lie detector test, various sensors are strapped onto the subject to record these physiologic changes. On the other hand, the pupil movements can be followed by a hidden T.V. camera with a telephoto lens. This noninvasive, secretive approach better suited military intelligence operations.[3]

Because a dilated pupil signals excitement, it also can signal the ignition of a sexual attraction when two people meet. Such eye changes were appreciated in the old Spanish court, where women would apply pupil enlarging eye drops to their own eyes. The medicine (belladonna, meaning beautiful woman), was atropine. One dose of this medication can keep the pupils dilated for almost a month. The name belladonna tells us the use of the drug. When a man meets a woman with dilated pupils, she appears more attractive and also gives the impression that she is attracted to him. This show of presumed interest by her might ignite a spark of mutual attraction in him, and help jump-start a relationship. This conclusion was put to a test by Professor Ekhard Hess.[4] When the two photos of the same face (pupils dilated in one photo, Figure 6.1) was shown to a series of experimental subjects, most of the subjects considered the face with the dilated pupil to be more attractive. Parenthetically, it is much easier to see the pupil in a person with blue eyes (i.e., there is more contrast with the black pupil against a light blue background than a brown background).

No one has ever offered a convincing functional explanation as to why the shade of the infant iris, be it blue or brown, is always lighter than that of the adult. One possible reason might be that the dark pupil is much easier to see against a light colored iris. Hence, emotionally induced pupil dilation of the infant would be more readily noticed by the mother. I was always amazed at the way my wife seemed to sense the innermost feelings of our babies when she held them. Along with their smiles and bouts of crying, she may have followed their emotional states by unconsciously reading their periods of adrenaline-induced pupil dilation.

As noted earlier, emotional pupillary dilation is a response to body adrenaline levels that reflects a level of heightened emotion (either positive, as joy, or negative, as pain or disgust). However, the pupil is not the only part of the eye that responds to endogenous adrenaline.

Eye changes from an elevation of the adrenaline level also include greater eye prominence (proptosis). This is due to adrenaline stimulation of involuntary muscles that raise the lid and push the eye forward (Muller's muscles of the lid and

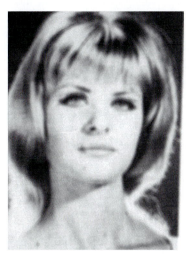

FIGURE 6.1 Two pictures of the same woman. In one photo the pupils are larger. Most observers thought the woman with the enlarged pupils was more attractive. (From Hess EH: *The Tell-Tale Eye.* Van Nostrand Reinhold, New York, 1975, with permission.)

orbital floor). Thus, the wide-eyed appearance of the screaming supporters seen at popular music concerts truly represent an adrenaline-induced enthusiasm.

Evolutionarily speaking, use of the pupil or its anatomic counterpart in registering emotions is not totally unique to the human. In many fish, the pupillary light response (which reduces the amount of light striking the retina in a bright environment) is taken over by a darkening of the cornea (somewhat like photochromic sunglasses). One fish, the guppy, shows a corneal darkening if threatened or vigorously handled by researchers.[5,23] Thus, this coupling of functions (i.e., light control and emotion display) has a long evolutionary heritage.

B. CRYING

As we worked with the pupillometer for the advertising agency, we found that one type of psychological stimuli consistently produced a large emotional pupillary response. That was the picture of a crying child.

Interestingly, tearing in response to an emotional stimuli is unique to humans.[6] For a moment, let's look a bit closer at the act of crying. Crying children can be heard as well as seen. However, if the crying is obscured by another noise (rain, thunder, river or ocean waves, village noise) a mother may not hear that her child is crying. How do we actually see that someone is crying? Tears appear on the cheek, and they strongly reflect light as tiny bright spots.* Thus, bright reflections on a child's cheek register crying. With crying, the lower tear meniscus gets higher. This produces a wider band of reflected light across the lower part of the eyes. Thus, crying produces two bright areas of light reflection (along the lower lid and on the cheek). These bright areas can be seen at distances up to 20 feet or more. In addition, the crying child is also an excited child, providing more clues to the mother.

Charles Darwin[7] reported that crying is a signal that the child is uncomfortable or in danger in every culture he studied. It is very useful as a survival response for the human infant and child. If the crying is not heard because of other masking noises or if the mother is deaf, the visual signs of crying are an important backup system.

C. EYE REDDENING

There is an acting troupe in Southern India in which the actor playing the role of the devil puts an irritating seed in his eyes, before the performance, to make them red. The red eye immediately tells the audience which actor is the devil. Is there a physiologic reason to associate red eyes with the devil?

We know that enraged, shouting people often show bulging, red eyes. Also, people working very actively in hot locations† perspire a great deal. The perspiration often drips into their eyes producing redness from the irritation. These two

* The short radius of curvature of a teardrop creates a very bright (virtual) reflected image.
† Hell is usually portrayed as having a very hot environment.

observations suggest that portraying the devil with red eyes makes both mytho-logic and physiologic sense. How does enragement or shouting produce red eyes?

Straining (as in lifting weights, fighting, or shouting) produces a Valsalva maneuver in which the venous return of blood from head and neck to heart is pre-vented by the bulging muscles of the neck. The muscle contraction forces large amounts of blood to back up into the head, producing a dilation of veins of the neck, face, orbit, and conjunctiva. This blood backup produces an instantaneous reddening of the eyes and face. At the same time, the eyes bulge because of the action of adrenaline on the orbital floor muscle, as well as the effect of the extra blood forced back into the orbital veins, which pushes the eyes forward. Tensing the muscles of the neck is frequently associated with rage. Therefore, rage can legitimately be interpreted in a person by the appearance of bulging red eyes.

D. EYE TO EYE CONTACT

There was a young and handsome bachelor who once worked in our clinic at the hospital. On Friday afternoons, we would celebrate the end of a week's work by going to a local pub. While we discussed the interesting cases of the week, he would unconsciously look at the young women who passed by. On occasion, he would make eye contact with one of them. In the event of mutual contact, he would excuse himself, go over to the woman, get her phone number, and return to our table. In time, it became part of a ritual that I anticipated and enjoyed. He later told me that, based on many encounters, he had learned that when the eye contact lasted longer than a second or two, the biologic attraction between the two was very strong. Scientific studies of eye movements while the subject observes vari-ous scenes reaffirm the fact that the eyes continually return to the objects of great-est interest.[8] Such studies have also showed that the number of glances, as well as the duration of the glance correlated well with the level of the subject's interest. In *About Faces,* the author T. Landau[9] notes anthropologist Irvin Devore's succinct observation "that if two people look at each other for more than six seconds, they will either make love or kill each other."

This eye contact phenomenon seems to be under subconscious control. Such involuntary eye contact does not only get triggered in cases of sexual attraction. At a cocktail party, such contact takes place whenever a person of importance (e.g., your boss or your biggest competitor), or someone you like, enters the room.

You may have noticed that many South American and Arab businessmen often wear sunglasses in a dark indoor setting. It has been suggested that the use of dark glasses could hide any clues such as inadvertent eye contact or unconscious pupil dilation, which might give away their innermost thoughts.

We also use our eyes to avoid potential confrontation. Professor E. Goffman[9] observed that at about 8 feet away, a stranger walking on the sidewalk will signal with his eyes which direction he or she will take to pass.

On the other hand, "street smart" people often suggest that when walking in an unfamiliar or dangerous neighborhood the safest demeanor is to keep your head up but don't catch anyone's eye. Finally, it is worth noting that most of us have difficulty talking to a person with a crossed eye. It is a bit distracting not to know with which eye you are in contact.

The British psychologist Simon Baron-Cohen[10] has suggested that we must be able to read the minds of others in order to become successful social creatures. He feels that our monitoring of the eye movements of others is very important in working with others.

A few years ago, I was interested in learning how precise we are at locking onto another's glance. Using an artificial eye centered on a protractor, which had a fine indicator line, I was able to lock onto the optical axis of the eye within less than half a degree. Such precision supports the importance given to the human ability to lock onto the glance of a fellow human.

E. BLINK RATE

Psychologist John Stern[11] noted that when President Nixon was televised answering the questions of reporters during the Watergate investigation, his voice remained calm, but his high blink rate gave away the fact that he was under great tension and was probably lying. Some experts have suggested that the increased sweating, which may have occurred when President Nixon lied, irritated his eyes and induced the increased blink rate. There might also be another reason why we increase our blink rate when under tension. Perhaps this reflex is related to the ancient fight or flight reflex, caused by a surge of endogenous adrenaline. During most fights, a great deal of dirt and dust are thrown into the air. This suggests that there is an advantage in keeping the eyes closed as much as possible (blinks) to prevent corneal abrasions that might be produced by the flying dirt and debris. Of course, one cannot fight successfully with the eyes closed. However, since the retinal image remains for a short while after eye closure (the afterimage), periodic blinks would not interfere with a sense of continuous action. Although supposition, it could explain how any tension- or anxiety-producing event will stimulate adrenaline release and increase the blink rate. It might also be interesting to review the old videotapes of Mr. Nixon to see if his pupils were dilated while he was interviewed.

This is another, albeit very rare, use of the blink in communication. It is the heartbreaking story* of a gifted writer, who suffered a brainstem cerebrovascular accident that rendered almost all his muscles paralyzed, except those of his neck

---

* Jean-Dominique Bàuby was the 43-year-old successful editor of a French high fashion magazine (Elle), when he suffered a devastating stroke. During his convalescence in a chronic disease hospital, he described his mind as taking flight like a butterfly in the prison of his own body, which he likens to a diving bell. The book is entitled *The Diving Bell and the Butterfly* (A.A. Knopf, New York, 1997).

and one eyelid. He was able to dictate the story of his experiences as a quadri-
plegic by having colleagues continually repeat a special alphabet, stopping them
with a blink when they had reached the proper letter. This arduous task resulted in
a book, which was published 2 days before his death.

F. POSITIONING OF BROWS AND LIDS

More than 100 years ago, Charles Darwin wrote that the facial expressions
of emotions are universal and not learned.[7] He based his hypothesis on observing the facial expressions in animals, human children, blind humans, and members of different cultures. Recent scientific studies tend to support these conclusions. Use of the eyes and lids are an important part of facial expression.[12–14] In
Figure 6.2, top left, the eyes are seen to be opened wide in surprise, with the
lower lids depressed, the upper lids elevated, and the brows arched. The upper
lid elevation exposes more white. This response is not limited to humans. Not
long ago, I visited a primate breeding facility in North Carolina. As I stared at
the rhesus monkeys moving freely in a wire fence enclosed corral, the leader
approached me and arched his upper lids, showing a large area of the white of
his eyes, much like the human surprise response. "He's warning you to keep
out," my guide told me.

In Figure 6.2, bottom right, anger is expressed by elevating the upper lids, let-
ting the lower lid just line up with the lower edge of the iris, and dropping and

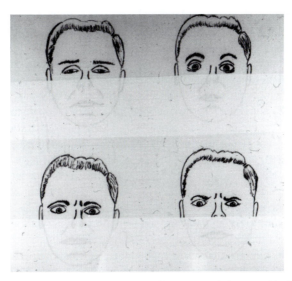

FIGURE 6.2 *Top right:* Sadness (a lowering of the brow and the eyes). *Top left:* Surprise (a
strong raising of the brow and upper lid). *Lower right:* Fear (a neutral brow and raising of the upper
lid). *Lower left:* Anger (a raising of the inner brow and raising of the upper lid).

wrinkling both inner brows. In Figure 6.2, bottom left, note that fear is similar to anger, but the inner brows do not dip down. On the other hand, sadness is usually conveyed with both a lowered brow and a lowering of the eyes themselves, as seen in Figure 6.2, top right.

What is quite surprising is the size of the vocabulary of eye signs. This seems to confirm the response of American film star George C. Scott when asked about the secret of his success as an actor. He said, "In acting … it's all in the eyes."

G. DETERMINING AGE THROUGH THE EYES

A few years ago a contact lens company was approached by a "beauty consultant" who had observed that the eyes of young women who appeared to have a dark ring at the edge of the colored iris look more attractive. In Figure 6.3A, in color insert, note the striking appearance of the eye of the young woman with the highlighted iris border. In Figure 6.3B, in color insert, note the eye of an older woman in which the iris border is covered by an arcus senilis. Most people over 40 have such a ring composed of either cholesterol (arcus senilis) or calcium (Vogt's limbal ring).[15] Could these phenomena be nature's device for announcing our ages?

SUMMARY

A. ADVANTAGES OF EYE COMMUNICATION

The subconscious eye signs described above suggest that prior to verbal language there had been an evolutionary advantage in one human signaling one's inner feelings and emotions to another human.

This rich vocabulary of eye communication of the emotional state further suggests that evolution has invested a great deal of effort in this regard. Aside from telling others in your family or community of your pain, sympathy, fear, ardor and so on, is there another advantage of signaling your emotional state to others? A number of years ago, Bronowski and Bellugi[16] suggested that most animals are pinned to the tyranny of the moment. That is, their emotional behavior is an instant, inherited, preprogrammed response. The human being is a bit different. We can be taught to modulate this instant emotional behavior by a form of reasoning. Many of us can detach ourselves and think about the moment. I'd like to suggest that one way of controlling an instant emotional response is to make use of your community. By announcing your emotional state with an eye sign or a facial gesture to those around you, these signs may stimulate others to put "the brakes" on your impending act of rage, anger, or contempt before it erupts. It might be useful for the reader to think of other ways that the ocular signaling of emotions or subconscious thoughts might help strengthen the bonds of a human community.

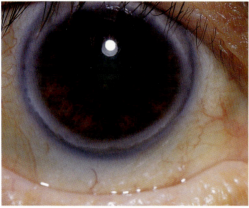

FIGURE 6.3 *Top:* Edge of iris is highlighted by a dark ring to make eye look more "attractive." (Courtesy of Pilkington, Barnes Hinds Corp. Sunnyvale, CA). *Bottom:* In older eyes, the apparent dark ring is obscured by a white ring of cholesterol (arcus senilis).

FIGURE 6.4 Painting of children with their large eyes showing great sadness by the artist Walter Keane. Keane wanted to show children on the threshold of life, facing a world of sadness. (Courtesy of Boston Public Library from Walter Keane, Tomorrow's Masters Series, Johnson & Meyer Publishing Company, Redwood, CA 1965.)

FIGURE 6.5 In these nine classic portraits from the past six centuries, one of the eyes is placed in the horizontal center of the frame. The white line marks the horizontal center. Portraits are by Roger von der Weyden (c. 1460), Sandro Botticelli (c. 1480), Leonardo daVinci (1505), Titian (Tiziano Vecelho, 1512), Peter Paul Rubens (1622), Rembrandt van Rijn (1659), Gilbert Stuart (c. 1796), as reproduced on the U.S. $1.00 bill, Graham Sutherland (1977), and Pablo Picasso (1937). (From Tyler, C: Eye placement in portraits. *Scope* (newsletter of the senior Ophthalmologist Interest Group, American Academy of Ophthalmology, Vol. III, Winter, 1997, p. 2, with permission.)

FIGURE 6.6 William Blake's painting, "The Evil Angel Orc versus the Good Angel Los." The eyes of the evil angel are featureless. (From Fontana D: *The Secret Language of Symbols*. Pavilion Books, London, 1993, with permission.)

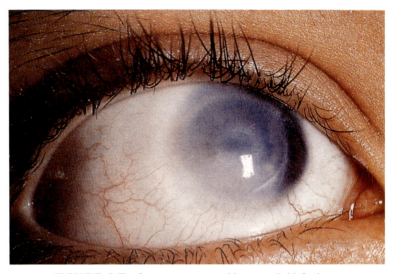

FIGURE 6.7 Opaque cornea caused by congenital infection.

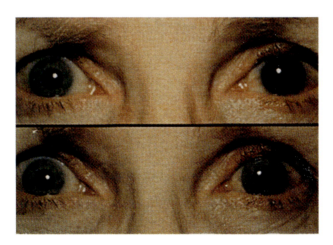

FIGURE 6.8 In upper photos, note absence of iris in left eye. In lower photos, patient looks normal with use of specially made cosmetic contact lenses. (Courtesy Leroy Meshel, M.D.)

FIGURE 6.9   The Automeris io or bull's eye moth of the Western hemisphere. (From Daccord M, Triberti P, Zanetti A: *Simon and Schuster's Guide to Butterflies and Moths.* Simon and Schuster, New York, 1988, with permission.)[22]

FIGURE 6.10   The Lasiommata megera or wall-brown butterfly of Europe and North Africa, which demonstrates two large eye spots with circular corneal highlights and six smaller eye spots along the lower wing edge. Note the smaller highlights on the smaller eyes. (From Caddordi M, Triberti P, Zannetti A: *Simon and Schuster's Guide to Butterflies and Moths.* Simon and Schuster, New York, 1988, Fig.84, with permission.)

FIGURE 6.11 The wolf spider has six eyes in different sizes and different corneal curvatures. Note that two of its six eyes have smaller radii of curvature and thus smaller corneal reflections. (From Caddordi M, Triberti P, Zannetti A: *Simon and Schuster's Guide to Butterflies and Moths.* Simon and Schuster, New York, 1988, p. 16, with permission.)

FIGURE 6.12 This ensemble of photos shows the multiple eyes of two spider species. It also shows a butterfly with six eye spots. Finally, it shows a spider feeding on a butterfly. (From Preston-Mafhem R, Preston-Mafhem K: *Spiders of the World.* Facts on File Publications, New York, 1984, pp. 118, 54, 11, with permission.)

FIGURE 6.13 Eye spots on buttocks of South American frog, Physalaemus. (From the cover photo of Owen D: *Camoflage and Symmetry. Survival in the Wild.* University of Chicago Press, 1984, with permission.)

FIGURE 6.14 Two eye spots on the rear section of a Tetradodon palembangensis. (From Axelrod HR: *Dr. Axelrod's Atlas of Freshwater Aquarium Fishes,* 2nd Ed. T.E.H. Publishers, Neptune City, NJ, with permission.)

FIGURE 6.16   The many eye spots on the wings of a peacock.

FIGURE 6.18 Note the black eye bar that goes from the eye to the mouth in H. burtoni. This area instantly turns dark when the male fish establishes its territory. (From Fernald RD: Vision and behavior in an African cichlid fish. *American Scientist* 72:58, 1984, with permission.)

FIGURE 6.20 The head statues of Easter Island **(A)** with and **(B)** without eyes (From Westwood J: *Atlas of Mysterious Places.* Marshall Editions Ltd., London, 1987, p. 132 and Picker F: *Rapa Nui, Easter Island.* Paddington Press Ltd., Two Continents Publishing, New York, with permission.)

B. EYE COMMUNICATION IN PAINTING

The uses of eyes to convey emotion is dramatically portrayed in the famous paintings of Walter Keane. The enlarged eyes of the children painted by Keane in Figure 6.4 in color insert, convey a sense of sadness and need in a way unmatched by any other technique. The enlarged eyes of the children in the paintings cry out for help as if the pictures included an audio track of actual crying.

Famous portrait artists have made considerable use of eye contact in their works. Dr. Christopher Tyler has measured the horizontal and vertical positions of the eyes in portraits by 170 such artists. As can be seen in a sampling in Figure 6.5 in color insert, most eyes were found to be placed in the geometric center of the portrait. One might speculate that placing the eyes centrally raised the probability of developing an eye to eye contact between subject and viewer, making the connection more personal.

C. DETERMINING HEALTH THROUGH THE EYES

In many cultures, people with eyes that are scarred of deformed in some way were referred to as evil eyes. In Figure 6.6 in color insert, the artist William Blake labels the evil angel with blank, featureless eyes. Might there be a human survival advantage in labeling these eyes as evil, and avoiding people with such eyes?

Medically speaking, eyes with opaque white corneas as in Figure 6.7 in color insert, (cause featureless looking eyes by covering iris and pupil) may be associated with congenital errors of metabolism, childhood vitamin deficiencies, and congenital infections such as syphilis. White pupils* (caused by cataracts) in people of childbearing age may also be associated with congenital errors of metabolism, severe childhood vitamin deficiencies, and congenital infections such as rubella.[17] Because of poor vision, these eyes often become crossed, making them look even more unsightly. Finally, people without an iris (aniridia, Figure 6.8 in color insert) often die at a young age from cancer of the kidney.

As implied above, people with these "deformed eye" conditions often have severe systemic illness either inherited or related to damaged organ systems. Might our avoidance of people with an "evil eye" be considered a social technique for avoiding people who might be unable to reproduce normal offspring or help defend a family or village?

Interestingly, people with a deformed eye, who wear a black patch to cover the eye[†] (such as former Israeli general Moshe Dyan) are considered dashing or heroic looking. Thus, having only one eye is not a reason to be shunned, but having a deformed eye is.

---

* White pupils (leukocoria) may also be caused by a retinal cancer (retinoblastoma), which is often inherited.

† Moshe Dyan, an Israeli military hero who lost orbital and eye tissue from an injury, avoided the danger of appearing to have a deformed eye by wearing a black patch.

2. ANIMAL EYE SPOTS

Of all the structures of the human body, why might the eye have been chosen for some of our nonverbal communication? I don't know. However, I do know that all along the evolutionary chain, eyes or replicas of eyes have been widely used for communication. Let's explore some interesting examples.

A. BUTTERFLY EYE SPOTS

The wings of the order Lepidoptera (butterflies and moths) demonstrate many interesting optical principles. Within the 165,000 species identified to date are combinations of colors and patterns that would impress the most sophisticated of art critics. The patterns and colors are achieved by the precise placement of both pigmented and nonpigmented scales that primarily take advantage of diffraction and scattering.* One of the most interesting patterns seen on butterfly wings are replicas of eyes.

What function might the "eye spot" pattern have? Experiments using birds and butterflies with eye spots have demonstrated that when the peacock butterfly, Nymphalis io, suddenly displays its eye spots by raising its hindwings and retracting the forewings, attacking birds draw back. When the eye spot pattern is rubbed off the butterfly wings by the experimenter, the same birds will attack and eat the butterflies.[18] Butterfly eye spots come in many configurations, almost all displaying a white spot on the center of the pupil that represents the corneal light reflection.

For example, the Automeris io or bull's eye moth of North and South America (Figure 6.9 in color insert) demonstrates two large eye spots when both wings are extended.

However, the story of eye spot patterns can get complicated. There are many butterflies that have six or eight eye spots. The wall-brown butterfly of Europe and North Africa (Figure 6.10 in color insert) has four on each wing. The odd thing is that the eye spots vary in size. The wall-brown butterfly shows one large eye spot on each forewing and three progressively smaller ones on each hindwing. What message or survival ploy might this represent? One answer came from one of my students after I had delivered a lecture on corneal optics. One of the slides that I use is a photo of a wolf spider, as seen in Figure 6.11 (see color insert). That spider has six eyes. Its top two eyes are larger than the bottom four. The top corneas have larger radii of curvature and produce larger reflections or highlights than the four smaller eyes. The student first asked me if spiders attack butterflies. Figure 6.12 (see color insert) answered his question by showing the capture of a butterfly by a spider. He then suggested that the butterflies with six or eight eye spots are trying to fool spiders with a similar number of eyes into think-

---

* A collector of preserved moths and butterflies observed that the wing patterns of night moths ultimately fade if left in the light. Daytime moths and butterflies have wing patterns that use diffraction and colorfast pigments.

ing that they are fellow spiders. On the other hand, the eight eye spots might also represent a dual strategy. The two large eye spots are used to fool birds, and the six smaller spots to trick spiders.[19]

B. EYE SPOTS IN OTHER SPECIES

The eye spot ploy is not unique to the butterfly. Figure 6.13 (see color insert) was striking enough to be used on the cover of a textbook concerning animal mimicry. It clearly shows the posterior aspect of a South American frog (an amphibian) demonstrating an eye spot replete with a central corneal reflection on each buttock. It seems very likely that a predator staring at this sight might sense that it represents part of the face of a larger animal and either hesitate before striking or actually withdraw.

Eye spots are also found near the tail section of many ocean fish, as in Figure 6.14 (see color insert) Reptiles also have eye spots. In Figure 6.15 the back of the head of a hooded cobra displays two dramatic eyes and a nose.

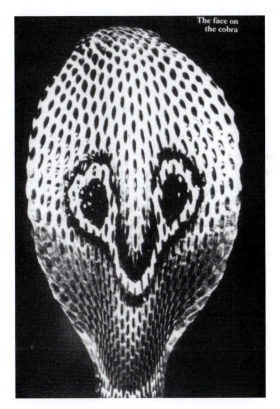

FIGURE 6.15 Back of the head of a hooded cobra with simulated eyes and nose (From Michael J: *Natural Likeness.* E.P. Dutton, New York, 1979, p. 59, with permission.)

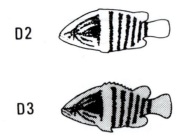

FIGURE 6.17 Drawings showing different presentations of the eye spot of the oyanirami. Only the model that uncovered the eye spot behind the real eye elicited a response by the other fish. (From Kohda Y, Watanabe M: The aggression releasing effect of the eye-like spot of the oyanirami Coreopera kawanebari, a freshwater serranid fish. *Ethology* 84:162–166, 1990, with permission.)

Now, let us look at the detail of one of the most beautiful eye spots in nature. The peacock shows many eye spots when it expands its wings (Figure 6.16 in color insert). Experts seem to feel that they are used more for sexual display than as a defensive strategy. However, I would think that the sight of its 80 or more eye spots looking down on a potential predator could well have a strongly disquieting effect on the predator.

The fresh water fish, oyanirami, uncovers a large eye spot behind its real eye when it starts to fight. Experiments have shown that uncovering of the eye spot is akin to announcing its desire to fight (Figure 6.17). Once defeated, the fish covers the eye spot, making it inconspicuous.[20]

Structures on or around the eyes of fish are also used for other social communication signals. Among the teleosts, the cichild fish species has an elaborate signaling system. In Lake Tanganyika there is a brightly colored cichild known as Haplochromis burtoni. The male can instantly induce a black bar of pigmentation (known as the eye bar) between the eye and mouth (Figure 6.18 in color insert). It neurogenically triggers this response when it has successfully acquired a territory, somewhat like a trophy or medal given to a successful human competitor. In the case of H. burtoni, the eye bar also announces that it is prepared to fight off any male invaders and that it is seeking a female. Since only about 10% of the males acquire territory, attract females, and mate, this signal is vital for both survival and reproductive success. Interestingly, eye spots are also found in the plant kingdom. The bark of this palm tree is a veritable array of eye spots (Figure 6.19), which may well frighten away creatures who would do harm to the tree.

However, the most spectacular show of eye aggression in the world is found in the horned lizard. Its eye not only gets red, but it squirts a stream of blood at its attacker. Naturalist Dr. Wade Sherbrooke told me that the horned lizard is capable of firing a cubic centimeter of blood up to a distance of 6 feet, and that it can reload, so to speak, and refire the blood squirt a few times. The animal can build up a volume of blood behind its eye, forcing its eye to bulge forward and the skin

FIGURE 6.19   The trunk of the palm tree seems covered with "eye spots."

around the eye to expand. Once the pressure builds up in the eye area, the blood is ejected from the capillaries of the conjunctiva (Dr. W. Sherbrooke, personal communication, Santa Fe, NM, 1995).

SUMMARY

A. THOUGHTS ON ANIMAL EYE PATTERNS

From some of these observations, one might draw the conclusion that nature commonly uses the eye spot to announce the presence of animal life. Why? There is nothing else that looks quite like it in the living world. In terms of compactness, it usually occupies less than 1% of the surface area of most animals. Being one of the most dramatic organs in terms of performance and importance to survival, it also has high recognition throughout the animal world. Finally, in further support of this argument is the common observation that some birds and monkeys will attack the eyes of an enemy as their primary offensive strategy.

B. THOUGHTS ON MAN-MADE EYE PATTERNS

The 30 foot statues of heads on Easter Island in the Pacific Ocean have baffled scientists for years. These eyeless heads seemed lifeless until someone located an odd white, elliptical plate on the ground next to a statue. Looking at Figure 6.20A and B in the color insert, it is easy to see that once these eye plates were replaced in the head, the statue took on lifelike qualities. Notice that the shiny material used to make the eyes also reflect light, producing corneal highlights that made them look even more lifelike.

In a sense, toy manufacturers of today took the lesson one step further. Mickey Mouse and Donald Duck toys with black felt eyes never sold well because they did not have the same lifelike properties as those with the shiny button eyes that produced a corneal reflection or black felt eyes with painted white corneal highlights. Just as in the butterfly world, an eye spot only looks real if a corneal highlight is present.

Toys and statues are not the only human vehicle for eye spots. When Nippon airlines discovered that birds were being sucked into the jet engines of their airplanes, they consulted a naturalist. The solution was eye spots, painted on the intake of each jet engine. Using this strategy, far fewer engines and birds were destroyed after placing the eye spots on the engines.

A story from Marlise Simons in the *New York Times* of September 5, 1989, described the successful use of a modified version of eye spots.

> For the moment, the Bengal tiger has met its match in the two-faced human.... In order to prevent tiger attacks from the rear to Indian workers, a student of the Science Club of Calcutta noted the protective eye spots on butterflies, beetles and caterpillars. Thus, he suggested that rural workers at risk wear face masks on the back of their heads. Happily, the ploy worked.

REFERENCES

1. Bower TG: *The Perceptual World of the Child.* Harvard University Press, Cambridge, MA, 1977.
2. Zuber B, Miller D: A simple inexpensive pupillometer. *Vis. Res.* 5:695, 1965.
3. Hess EH: *The Tell Tale Eye.* Van Nostrand Reinhold, Inc., New York, 1975.
4. Goffman E: *The Presentation of Self in Everyday Life.* Doubleday, Garden City, New York, 1959.
5. Iga T, Takabatake I, Watanabe S: Nervous regulation of motile iridophores of a fresh water goby Odontubutis obscura. *Comp. Biochem. Physiol.* 88C:319–324, 1987.
6. Walsh FB, Hoyt MF: *Clinical Neuro-Ophthalmology,* Vol. 1. Williams & Wilkins C., Baltimore, 1969, p. 553.
7. Darwin C: *The Expression of the Emotions in Man and Animals,* John Murray, London; facsimile reproduction with an introduction by Konrad Lorenz. University of Chicago Press, Chicago, 1965. (Originally published 1872).
8. Argyle M, Cook M: *Gaze and Mutual Gaze,* Cambridge University Press, Cambridge, England, 1976.
9. Landau T: *About Faces.* Anchor Books, New York, 1989.
10. Baron-Cohen S: The Language of the Eyes. In *Mindblindness, An Essay on Autism and Theory of Mind.* MIT Press, Cambridge, MA, 1995.
11. Vogel S: In the blink of the eye. *Discover,* Feb 1989, pp. 62–64.

12. Ekman P, Friesen WV: *Unmasking the Face.* Prentice-Hall, Englewood Cliffs, New Jersey, 1975.
13. Ekman P: About brows; emotion and conversational signals. In *Human Ethology.* (Von Cranach M et al, eds.). Cambridge University Press, Cambridge and New York, 1979
14. Marsh P (ed): *Eye to Eye.* Salem House, Topsfield, MA, 1988.
15. Smolin G: Dystrophies and Degenerations in the Cornea. In *The Cornea* (Smolin G, Thoft RA, eds.). Little Brown, Boston, 1983, p. 334.
16. Bronowski JS, Bellugi U: Language, name and concept. *Science.* 168:699, 1970.
17. Mausoff FA (ed): *The Eye and Systemic Disease.* C.V. Mosby, St. Louis, 1975.
18. Best AD: The function of eye spot patterns in the Lepidoptera. *Behavior.* II:209–256, 1957.
19. Miller D: Butterfly eye spots and corneal reflections. *Eur. J. Implant. Refract. Surg.* 3:279, 1991.
20. Kohda Y, Watanabe M: The aggression releasing effect of the eye-like spot of the oyanirami Coreopera Kawamebari, a freshwater serranid fish. *Ethology* 84:162–166, 1990.
21. Blakemore C (ed): *Vision Coding and Efficiency.* Cambridge University Press, Cambridge, England, 1990, p. XV.
22. Daciordi M, Tribert P, Zanetti A: *Simon and Schuster's Guide to Butterflies and Moths.* Simon and Schuster, New York, 1988.
23. Shand J, Partridge JC, Lythgoe JH: Catecholamine induced colour changes in the corneal iridophores of the sandgoby. *Comp. Biochem. Physiol.* 94C:351–355, 1989.

PART

II

THE VISUAL BRAIN

7

CREATING VISUAL STORIES AND ILLUSIONS AROUND THE RETINAL IMAGE

Introduction
   A. History
   B. Brain as Story Teller
   C. Brain Processing in Animals
1. Visual Illusions
   A. Background
   B. Registering Human Faces
   C. Illusions that Emphasize Important Objects
   D. Time Compensating Illusions
   E. Stereoscopic Illusions
   F. Ambiguous Figures
   G. Apparent Motion
   H. Disappearing Images
   I. Global Impressions Come First
Summary
   A. Subjective Aspects of Visual Illusion
   B. Global Impression versus Detail Detection
   C. Facial Recognition

INTRODUCTION

The scientists who are interested in the subject of brain processing of the retinal image have polysyllabic titles such as neuro-ophthalmologists, neurophysiologists, neuroanatomists, neurobiologists, psychophysicists, and computer modeling experts. They use terms like veridical perception, pyramidal algorithm, iconic bottle necks, and isodipole texture. Their combined findings have been insightful, often brilliant, but difficult to understand because of technical language and the absence of discussion of the application to daily life. David Marr, in his book on computational vision,[1] described this absence of a practical framework colorfully when he wrote, "Trying to understand perception by studying only neurons is like trying to understand bird flight by studying only bird feathers." Friedrich Cramer, in his book, *Chaos and Order*[2] suggests that anatomic descriptions of the brain are similar to a railroad schedule. You get an exact description without strategy or reason for existence. Hopefully, the next four chapters will describe the different ways that the brain processes the information from the retinal image in a user-friendly manner, with speculation as to the utility of such processing. As in previous chapters, the theme of survival benefits will hover over many of the examples of visual processing.

A. HISTORY

Let's turn to the beginnings of understanding brain processing of the retinal image.

In 1619, a Jesuit priest by the name of Father Christof Scheiner wanted to find out what the retinal image actually looked like.[3] He used the eye of a freshly killed animal and pointed it toward a candle. To observe the image of the candle on the retina of the eye from the outside, he had to peel away the scleral outer covering of the eye so as to be able to see the image through the outer surface of the retina. The image of the candle was upside down! Yet, to all of us the world and all it contains looks right side up. Obviously, some processing of the retinal image must be performed before "vision" takes place.

About 100 years ago, an experiment on living people was performed that helped elaborate Father Scheiner's findings.[4,5] A group of experimental subjects wore special prism spectacles that literally turned the world upside down. Initially, this inverted world made life very confusing. However, after a few days of wearing the prism spectacles, everything appeared right side up again. This was the result of the visual processing in the brain.

Figure 7.1 is a cross-section of the head of a human embryo just a few weeks after conception. Note that the eyes are an outgrowth of the brain. As the baby grows, the eyes will move farther from the brain, and the connections of nerve fibers will stretch out and thin a bit. Nevertheless, the lesson from embryology is clear. The eye starts out as part of the brain, and retains its connection to the brain throughout life.

How large is the brain's contribution to vision in comparison to the eye? One way to answer that question is to compare the area of tissue in the fovea with the area of brain tissue that processes the foveal image. Dr. Jonathan Horton of the University of California[6] studied magnetic resonance images (MRIs) of a series

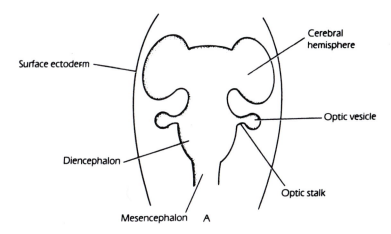

Surface ectoderm

Cerebral hemisphere

Optic vesicle

Diencephalon

Optic stalk

Mesencephalon   A

FIGURE 7.1   Cross-section of the head of a human embryo showing the eyes to be outgrowths of the brain. (From Shell RS, Lemp MA: *Clinical Anatomy of the Eye.* Blackwell Scientific, Boston, 1989, Figure 1.1, with permission.)

of patients with brain lesions that affected different parts of their field of vision. He and his colleagues were able to correlate the area of the damage with the size of the scotoma. Combining all their measurements, they concluded that the area of the brain that processes the foveal image is 1,000 times larger than the area of the fovea itself, which is less than 1 mm.[2]

B. BRAIN AS STORY TELLER

The famous physiologist Vernon B. Mountcastle[7] presented the idea that "the brain is a *story teller....* It is never completely trustworthy, allowing distortions of quality and measure, yet giving an order. ... Each image is conjoined with genetic and stored experiential information that makes each of us uniquely private..." Thus a perception is a story about an event.

A more sophisticated view of perception would see it as a flexible story, somewhat like a hypothesis with a certain probability of accuracy. On receiving the retinal image, the brain assigns a temporary story (hypothesis) that could quickly change when new information is provided.

In his book *Art and Illusion,* Gombrich[8] suggests that Leonardo gave his Mona Lisa a slight head turn and a veiled smile in order to produce a tentativeness that makes the face seem to change with each observation. Thus one might assume that Leonardo understood this concept of perception as a tentative hypothesis process and realized that a lighting change, a change in expectation, or a change in the mood of the observer might work to continually alter the impression of his painting.

Now let us look at a patient who has lost the ability to weave a story or a hypothesis around a retinal image. The neurologist and author Oliver Sacks[9] described such a patient, who had a serious cortical lesion. During the examination, Dr. Sacks showed the patient (Dr. P.) a glove:

> "What is this?" I asked holding up a glove.
>
> "May I examine it?" he asked, and taking it from me, he proceeded to examine it.
>
> "A curious surface," he announced at last, "infolded on itself. It appears to have," he hesitated, "five outpouchings if this is the word?"
>
> "Yes," I said cautiously, "You have given me a description. Now tell me what it is."
>
> "A container of some sort?"
>
> "Yes," I said, "and what would it contain?"
>
> "It would contain its contents," said Dr. P. with a laugh. "There are many possibilities. It could be a change purse for example for coins of five sizes. It could…"
>
> I interrupted the barmy flow, "Does it not look familiar? Do you think it might contain, might fit part of your body?"
>
> No light of recognition dawned on his face. No child would have the power to see and speak of continuous surfaces … enfolded on itself, but any child, any infant would immediately know a glove as a glove.
>
> The patient saw, as a computer machine might see; i.e., abstract key features and schematic relationships.

Unhappily, Dr. Sacks' patient could see *no meaning* or story in what he saw; he *could make no judgments about it* and *he had no emotional feeling toward it*. The part of the visual brain that interprets retinal images had been destroyed, yet he probably had 20/20 visual acuity. This part of the brain is the essence of human vision. It converts physical reality, known to Dr. P., into meaningful visual reality—a reality unique to each person and culture. It is a process known as *cognition*.

C. BRAIN PROCESSING IN ANIMALS

Throughout the animal kingdom, there is evidence that animals do not simply respond to the retinal image alone, but to a brain processed version of that image. After all, the retinal image is upside down, flat, and constantly vibrating due to a fine eye tremor. Yet, the animal reacts to the world as it were right side up, three-dimensional, and steady. Professor Jerry Lettvin of MIT[10] and his colleagues did a study on frog vision some years ago, which they entitled, "What the frog eye tells the frog brain." Examining recordings from the living frog brain, they noted a difference when the frog looked at a moving insect versus a stationary one. Obviously, the retinal image was the same in both cases. One might interpret this to mean that since the frog is a very practical creature with a limited brain volume, a stationary image of an insect might imply that it was dead, possibly poisoned, or not a real insect at all. A moving insect was a safer, surer meal. There have been other experiments that demonstrate that the frog also uses its limited color vision in a practical way. If you place a piece of blue paper and a green piece of paper on either side of a frog and then startle it, the frog will almost always jump onto the blue paper. I assume that the blue paper represents water, and therefore, a safe haven for a startled frog. What is the frog's behavior telling us?

Incorporated into the process of frog vision is some sort of screening analysis, that is, is the object food, is there danger? Once the question is answered, an action is taken. The frog brain, therefore, attaches a very personal meaning to the retinal image. Perhaps that is the story Professor Mountcastle had in mind.

1. VISUAL ILLUSIONS

A. BACKGROUND

There is a group of visual phenomena that seem to epitomize this relationship between the retinal image and vision. They are visual illusions. Illusions can be thought of as bogus stories of things that exist in the real world. A person looks taller than he really is, a face is seen in a cloud formation. Would a patient like Dr. P. see such illusions? There was a patient reported in 1918 who resembled Dr. P. The neurologists Goldstein and Gelb[11] wrote a 142-page article in which they tried to describe the patient's response to a myriad of visual stimuli. Interestingly, the patient was not fooled (i.e., did not appreciate a number of illusory figures). The vision of this patient would suggest that the process of appreciating an illusion may be closely tied to one of the essential aspects of vision. As you might expect, this point of view is not shared by all visual psychologists. Many take illusions to represent fascinating glitches or miswiring in visual processing. Others see illusions as examples of an algorithm in their computer model of the brain overriding a second algorithm.

On the other hand, Geothe observed,[12] "optical illusion is optical truth." Illusions are like visual surprises. They appear to contradict physical reality. Therefore, it would seem natural to dismiss illusions. However, they do display some basic biologic features. For example, you cannot override an illusion even if you know it does not make physical sense. I have projected illusions onto a screen in front of a room full of doctors and told them that all the lines in the illusion were really the same size. Yet, to every last member of the audience, one line continued to look longer than the rest.

Second,[13] one can still see an illusion, even if it is only exposed for a fraction of a second.* From this observation, we might conclude that they are like automatic reflexes (i.e., vital to survival).

Third, an appreciation of illusions stretches across almost all races and cultures, although it is true that some cultures will see one illusion and not another or see one illusion in a more exaggerated form than another.[13,14]

---

* An illusion can be appreciated if exposed for 50 to 150 msec.

FIGURE 7.2   A pattern within the leaves produces the strong impression of a human figure. (From Miller D: Brain processing of optical input: The perception of visual reality. *M.D. Computing* 11:36, 1994, with permission.)

Fourth, most illusions can still be appreciated even if the retinal image is made blurry.[13,14] From an evolutionary standpoint, this aspect of robustness suggests that illusions may be important for survival.

Fifth, illusions are visually appreciated from childhood to old age. A psychological response that remains for a lifetime can either be a glitch built into the wiring that never disappears or it can be very important to survival.

Finally, there are situations in which the illusion of an object has a stronger impact than the real objects in the scene. In Figure 7.2 the illusion of a human figure almost jumps out of the framework of leaves.

B. REGISTERING HUMAN FACES

There are many surprises connected to our recognition of human faces. therefore, in this section, we will use the term *visual surprise* instead of visual illusion. For example, we can identify the face of a celebrity from only a few features. Dr. Eli Peli and his research group at Harvard[15] took photographs of faces and then rendered them unrecognizable by electronically filtering out most details. Then they started replacing the larger, grosser facial features.* In Figure 7.3 we can see

---

* The size of an eyebrow is between two and four cycles per degree at a conversational viewing distance. Objects of this size are enhanced by the brain. This means that the contrast of objects of two to four cycles per degree in size are easier to see than the laws of optics would predict. This conforms to the peak seen in a typical contrast sensitivity graph. One could speculate that recognition of facial features has been given a high priority by the neural processing system.

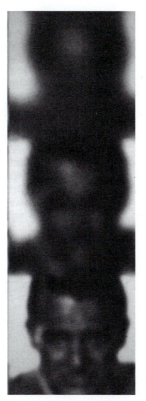

FIGURE 7.3   As more features are revealed (i.e., information from the high spatial frequencies), we ultimately recognize the actor Cary Grant. (From Peli E: Contrast in complex images. *J. Opt. Soc. Am. A* 7(10):2036, 1990, with permission.)

an example of what they did. Once certain large structures (the thickness of the brow or the width of the nose) were added, the face of actor Cary Grant becomes recognizable, although it is still quite blurred. Indeed, in general, we identify faces by cataloging the large features with much less regard for the nuances.

Figure 7.4 is a well-known portrait of William Shakespeare. It seems to look all right. Look again. Did you notice that the artist gave Shakespeare two right eyes?* Now look at Figure 7.5 (in color insert). Surprisingly, the Mona Lisa has a chalazion in the left lower lid. Had you ever noticed that before? You might ask just how small must a facial detail be to be commonly overlooked. In Figure 7.6 we see a self-portrait of the famous artist Albrecht Durer. It is not obvious that one side of his face is different than the other. Professor Friedrich Cramer[2] has recon-

---

* Some scholars argue that through this portrait (two right eyes), the artist tells us that Shakespeare was (akin to an artificial concept) not the real author of the famous plays.

FIGURE 7.4 The famous engraving of William Shakespeare by Martin Droeshout. The presence of two right eyes and a tunic put on backwards prompt some scholars to think that this was a disguised portrait of the real author of Shakespeare's plays, Edward de Vere, the 17th Earl of Oxford.

structed Durer's face with a computer so that a face is formed from the two right halves. The same was done with the left side of Durer's face. It is now obvious that Durer's face is not symmetrical. Professor Cramer's experiment teaches us that most of us do not notice the subtlety of facial asymmetry. Of course, the very careful portrait artist does capture the details. In Figure 7.7 (see color insert) we have rendered the shadows of a face into three different colors. Although the shadow is the same size and shape in all three cases, only when the shadow is black can we recognize the subject.*

* Dark shadows associated with faces is an example of learned behavior. Admittedly, one can learn to identify faces with white shadows. In the early days of television, news reels were shot on cine film. Time was saved when the film editors learned to read negatives (i.e., dark shadows looked white). (Hess EH: "Imprinting" in a natural laboratory. *Scientific American.* 227:24–31, 1972.)

FIGURE 7.6 A self-portrait of the artist Albrecht Durer, which appears in the Alte Dinakothekim Munich. *Left:* Normal reproduction. *Top right:* A face formed from the two right halves. *Bottom right:* A face formed from two left halves. (From Cramer F: *Chaos and Order.* VCH Publishers, New York, p. 153, 1993, with permission.)

If the picture of a face is covered with distractions, much as "snow static" on the T.V. screen, many people with brain disease will not be able to recognize the face. However, the normal observer will be able to discern the presence of the face of an infant through the cover of distracting noise in Figure 7.8 (see color insert). Distraction not only comes from visual noise covering a face. You might consider all the faces in a crowd as distractions that interfere with recognizing an unusual face. How quickly can you identify the face with a different feature in Figure 7.9? In two of the three scenes, a new feature has been added to one face in the crowd, while in the third scene a feature has been subtracted from a face. To some of us, such small changes will jump out. In general, fewer of us will notice the absence of an expected feature than the presence of an added one. This message was highlighted in a story ("The Adventures of Silver Blaze") by A. Conan Doyle in which his keenly observant detective is the only one to note that the watch dog did not bark on the night of the murder. Holmes realizes that the absence of barking meant that the dog knew the murderer.

A study by Benson and Perrett[16] demonstrated that subjects responded 50% more quickly to a caricature of a person than a photograph. What might be the reason? Look at Figure 7.10, in which the famous *New York Times* caricature artist, AI Hirshfeld, has drawn the faces of entertainers Sammy Davis Jr., Humphrey Bogart, Judy Garland, Laurel and Hardy, John Wayne, and Katharine

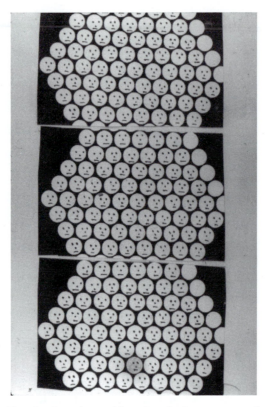

FIGURE 7.9 Three drawings of a sea of faces. In each of two scenes, one face has an additional feature. In the third scene, one face has one feature absent.

Hepburn. Each is recognizable, even though you may never have seen the caricature before. We actually file faces in our brains in the form of line drawings, in which prominent or unusual aspects are given higher priority, that is, a good caricature sketch should accelerate face detection. Are some of us better than others at face recognition?

Grüsser et al. have reported[17] that female subjects (from adolescence onward) are remarkably better than male subjects (of the same age) in face recognition. On the other hand, schizophrenic patients have impaired face recognition.[18]

In sum, we are surprisingly skillful at identifying faces even if the retinal image of the face presented is in a partial, blurred, or distorted form. We can assign a story to a face that presents only a minimum of information, such as a line drawing with no texture. On the other hand, we often overlook certain facial detail in generating a story about a face. Perhaps the benefit of such a system is to minimize the space in our brain that stores memory data. Although the primate

FIGURE 7.10 Al Hirshfeld's caricature renditions of entertainers: Sammy Davis Jr., Humphrey Bogart, Judy Garland, Laurel and Hardy, John Wayne, and Katharine Hepburn. (From Hirshfeld A: *Hirshfeld by Hirshfeld*. Dodd, Mead, New York, 1979, p. 195, with permission.)

brain possesses a facial recognition center, we assume, on the basis of the illusions presented, that the human is probably more skilled than the monkey at recognizing a larger library of faces and facial expressions.

C. ILLUSIONS THAT EMPHASIZE IMPORTANT OBJECTS

1. Verticals

A standing enemy presents a greater danger than one seated or kneeling. This observation is even recognized throughout the animal kingdom, where a standing bear or gorilla or an erect cobra is a clear sign that an aggressive act will follow. A vertical human figure is also potentially more dangerous than a seated one. In this section, a number of visual illusions will be presented that exaggerate the vertical form. Such magnification might be considered a device of the visual processing system, which reflexly highlights a potential danger. Figure 7.11 represents two arrows forming an inverted "T". Although the horizontal and vertical bars are the same length, the vertical bar looks larger to observers in all cultures. In fact, in a study in which a similar figure was shown to members of 13 African tribes as well as a group of Americans living in the Chicago area,[19] the African subjects saw the

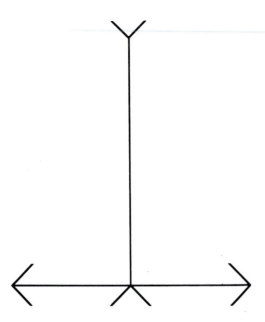

FIGURE 7.11   Two arrows forming an inverted "T". Although the bars of the arrows are the same size, the vertical bar looks longer than the horizontal bar.

vertical bar more exaggerated than the Americans.* Note that the illusion remains, even when the two bars are blurred. One might mentally project a standing human figure hovering over a fallen human figure of the same size to appreciate what survival purpose might be served by mentally enlarging the vertical figure. Let's expand this notion. In Figure 7.12 (see color insert) are two vertical bars sitting on a horizontal line. Suppose for a moment that they represent two creatures on the branch of a tree. The second diverging line represents another branch coming off the trunk of a tree. Of the two vertical figures, the heavier one would most likely be situated closest to the trunk (shorter lever arm, greater branch thickness) so as not to break off the branch. In this illusion, the vertical form closest to the trunk looks larger. This could be an adaptation by the visual system to highlight the larger (i.e., heavier, more realist danger). Note the effect is present even if the figure is in a fog or a blur. Figure 7.13A depicts the well-known Muller-Lyer illusion. The vertical line connected to the diverging lines looks longer than the vertical line of the same dimension connected to the converging

---

* Professor Gregory has suggested that most illusions have features that are related to the depth clues we get from our modern sense of perspective. He feels that we in the "carpentered world," and see certain parts of a figure as if it were a three-dimensional perspective drawing. In that context, certain lines that look far away are compensated for and look larger. (Gregory RL: *How The Eyes Deceive The Listener* 68:15–16, 1962.)

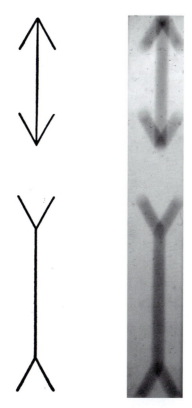

FIGURE 7.13 **(A)** The Muller-Lyer illusion. **(B)** Even when blurred, the one connected to the flaring lines looks longer.

lines. This illusion persists even when blurred (Figure 7.13B). Now let us add some "flesh and blood" to the illusion by creating stick figures. In Figure 7.14 (see color insert) the figure standing with arms and legs spread apart looks much larger than the hunched figure, yet the size of their vertical torsos is the same. Once again, the illusion persists even in a fog.

It would appear that the visual brain plays tricks on us by exaggerating the vertical. I have tried to suggest that such visual processing may be for our own good, that is, the brain has added a practical meaning to the retinal image, by identifying a potential danger.

2. Distant Objects

As a road leads off into the distance, its parallel sides seem to converge. This impression is the result of the laws of optics. The retinal image of the road before us is larger than the distant road. The retinal image of the entire road is a continuous merging of larger into smaller images, giving the sides of the road a converging character. Expecting that the converging appearance of a road signifies a long

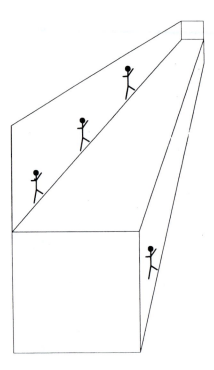

FIGURE 7.15   The height of the three figures positioned along the receding stage are exactly the same.

distance can produce interesting illusions. Can we reconstruct the illusion so that it might represent some survival value? In Figure 7.15 are three stick figures drawn along a stage with converging sides in order to exaggerate the feeling of "far away." The three stick figures are really all of the same size. However, the third figure looks much larger. Why? Two learned biases conflict here. The third figure should be farthest away because of the effect of the converging lines. Under normal circumstances, that figure would then produce a small retinal image. The brain process of size constancy (which will be discussed later) would then make that figure look larger. But in the drawing the figure is already made as large as the other two. Therefore, the process of size constancy can do nothing but make the third figure appear as a giant. Our expectations have played a trick on us. How strong is this illusion? Would it exist under unusual or adverse conditions? In Figure 7.16 (see color insert), all the stick figures are drawn to the same size, yet the illusion persists whether the figures are turned upside down or placed in a fog. It would seem that if the visual system is presented with clues that place an object or a person far away, it can introduce a telephoto effect that enlarges the distant object, without the use of enlarging optics. Does such a mechanism give the human being an advantage? It probably does in certain environments. If we are on a mountain and look down into the valley, we will see people and animals moving. We can identify them as people and animals although their retinal images present them as little more than tiny dots. However, because we know them to be people and animals from past experience, we enlarge them in our minds and assign certain characteristics to them (i.e., we surround them with a story). Psychologists call this phenomenon *size constancy.* We assign a certain size to the figures even though their retinal images are small.*

There are situations in which the size constancy mechanism seems to be disconnected. Studies have shown that long jump runners adjust their stride by using the real size of the retinal image to monitor time to impact.[20]

The same examples of size constancy are not found in all cultures.[13,14] The following experience bears out this assumption. Anthropologist Colin Turnbull lived with the Pygmy people of the Ituri Forest in the Congo in the late 1950s. He wrote a book chronicling his experiences, *The Forest People.*[21] In it he describes an event during which he and his Pygmy friend, Kenge, go on a trip out of the forest and into the Ishango National Park. Kenge had never been outside of the thick, dense forest in which the Pygmy people live. The following is Turnbull's description of Kenge's perceptual confusion with sizes and distances when he leaves the forest.

> When Kenge topped the rise, he stopped dead. Even the smallest sign of mirth suddenly left his face. He opened his mouth but could say nothing. He moved his head and eyes slowly and unbelievingly. Down below us, on the far side of the hill, stretched mile after mile of rolling grasslands, a lush, fresh green, with an occasional shrub or tree standing out like a sentinel into a sky that had become brilliantly clear. And beyond the grasslands was Lake Edward—a huge expanse of water disappearing into the distance, a river without banks, without end. It was like nothing Kenge had ever seen before. The largest stretch of water in his experience was what he had seen when he stood, like Stanley, at the confluence of the Lenda and the Ituri. On the plains, animals were grazing everywhere—a small herd of elephant to the left, about twenty antelopes staring curiously at us from straight ahead, and down to the right a gigantic herd of about a hundred and fifty buffalo. "What insects are those?"
>
> At first I hardly understood; then I realized that in the forest the range of vision is so limited that there is no great need to make an automatic allowance for distance when judging size. Out here in the plains, however, Kenge was looking for the first time over apparently unending miles of unfamiliar grasslands, with not a tree worth the name to give him any basis for comparison. The same thing happened later on when I pointed out a boat in the middle of the lake. It was a large fishing boat with a number of people in it, but Kenge at first refused to believe this. He thought it was a floating piece of wood.
>
> When I told Kenge that the insects were buffalo, he roared with laughter and told me not to tell such stupid lies. When Henri, who was thoroughly puzzled, told him the same thing and explained that visitors to the park had to have a guide with them at all times because there were so many dangerous animals, Kenge still did not believe, but he strained

---

* Butterfly wings often have patterns of vertebrate eyes or heads of predators. These are considered deceptive devices that might frighten away butterfly predators. The fact that these patterns are much smaller than the parts they imitate suggests that the butterfly predators may not have "size constancy."

his eyes to see more clearly and asked what kind of buffalo were so small. I told him they were sometimes nearly twice the size of a forest buffalo, and he shrugged his shoulders and said he would not be standing out there in the open if they were. I tried telling him they were possibly as far away as Epulu to the village of Kopu, beyond Eboyo. He began scraping the mud off his arms and legs, no longer interested in such fantasies.

There is an interesting flip side to this story. Black Africans living in the plains are very familiar with size constancy, and make use of it all the time in real life. However, they have difficulty transferring the concept to a picture or photo of real life. In the late 1950s, the Rayleigh bicycle company advertised their bicycles in Africa. A very popular advertisement showed a black man on a Rayleigh bicycle escaping from a lion, seen in the distance. The overwhelming response to the ad from the African black community was that it made no sense. Why should the man on the bicycle be trying to escape from such a tiny lion? Recall that these people lived on the plain, as opposed to the pygmies described by Colin Turnbull, and understood that, in real life, a small distant animal was really of normal size. However, they did not transfer size constancy to the picture.

There are other clues besides converging lines that suggest that something is far away or close by. For example, in Figure 7.17 (see color insert), are two sets of five balls.[22] The center ball in each figure is the same size. However, the central ball surrounded by small balls appears larger than the central ball surrounded by large balls. If the theory that our visual system can learn to enlarge the impression of distant objects is correct, than the four smaller balls suggest that they are farther away than the four larger balls. Our visual brain zooms up the central ball, which is surrounded by four smaller ones (size constancy) possibly highlighting its approach toward us.

3. The Moon Illusion

The moon overhead only subtends an angle of about one degree and looks relatively small. However, when the moon drops to the horizon, it appears larger. Experiments have shown that the moon at the horizon appears on average at least 1.5 times larger than the moon in the sky.[14] Many scientists have tried to explain this apparent magnification by invoking the concept of atmospheric optics. Since light rays, emanating from the moon on the horizon, travel through a slightly denser atmosphere, some have thought that the moon's enlargement was produced optically by an atmospheric magnification. However, scientific experiments show that optics cannot explain it and that the illusion only occurs in the mind of the beholder. It is the human brain that takes the small retinal image of the moon and creates a magnified impression. In Figure 7.18 (see color insert), I have tried to simulate the moon illusion using the distance clues mentioned previously. The moon at the horizon is flanked by earthly clues (the converging lines of the river and the valley between two mountains) and looks larger than the moon up in the sky or the moon in the foreground (all moons measure the same size).*

---

* It is possible that aside from the earthly clues that there is also a "moon on the horizon enlarging mechanism" in the visual system.

Therefore, the brain mechanisms involving converging lines and size constancy join to give us the impression of a larger moon (or sun since they both subtend the same angle) when either is close to the horizon. It is easy to understand how certain pre-industrial tribes might give supernatural significance to this illusion.[23]

However, there may be a more practical theory to explain the advantage of the moon illusion. If bringing attention to an object through magnification is the function of a visual illusion, then the moon illusion can be thought to act like an alarm clock for fishermen. Ocean waters contain more fish during the high tides.[32] Tides are highest when the moon is closest. Therefore, the enlarged moon or sun on the horizon might function to summon fishermen, as in Figure 7.19 (see color insert).

D. TIME COMPENSATING ILLUSIONS

In their book *Hitting Blind,* ophthalmologists Harold Stein and Bernard Slatt[24] present pictures of many professional tennis stars hitting the ball without looking at their racquet as contact is made with the ball (Figure 7.20) (thus, the title *Hitting Blind*). Dr. Stein feels that the tennis pros hit the ball at such high speeds that there is not enough time to visually track the ball and then visually control the

FIGURE 7.20 Four professional tennis players, not looking at the ball during contact (From Stein H, Slatt B: *Hitting Blind,* Dons Mills, Ontario, Musson, 1981, with permission.)

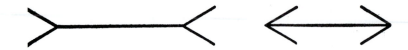

FIGURE 7.21   A horizontal Muller-Lyer type illusion similar to Figure 13.

motor response necessary for proper ball contact. The players in the picture must anticipate where the ball will bounce from watching it leave their opponent's racquet. Once their brain predicts (i.e., creates a story as to where the ball will bounce), their neuromuscular system starts preparing to hit the ball at that location. Further eye tracking is not needed. Unfortunately, few of us can perform at their level of accuracy. I use the word "illusion" to describe the way that these athletes prepare to hit a brain simulation of the ball. Talented athletes are better at creating this illusion than the rest of us.

In baseball, good fast ball hitters do the same thing. They must unconsciously predict where the ball will cross the plate just as the ball leaves the pitcher's hand. Mel Ott, one of the great hitters of the 1930s, actually articulated this phenomenon when he suggested that watching the spinning seam of the ball as it left the pitcher's hand helped him predict the position of the ball as it crossed home plate. On the other hand, very successful pitchers are the masters of "late movement," that is, a pitch that dances away from the batter after the bat is already in motion.

I would like to propose that there is also another visual device that we use to follow certain fast moving animals. The device is based on the horizontal version of the aforementioned Muller-Lyer illusion in Figure 7.21, in which the horizontal arrow with the diverging segments looks longer than the horizontal arrow with converging segments, although both horizontal lines are the same length. A related illusion, shown in Figure 7.22A (top), demonstrates that the view of a lunging animal looks longer than one of the same size in a crouching position. In the bottom of Figure 7.22B, the presence of paw prints in segment BC makes it look longer than segment AB (without prints), although both segments are the same length. Admittedly, Figure 7.22B is distinctly related to Figure 21. How might these visual illusion mechanisms help the primitive hunter? Suppose the hunter is trying to hit a moving animal with a rock or stick. If the animal is moving too fast, then the hunter has a problem similar to the fast ball hitter. By overestimating the size of the outstretched form of the animal or the tracks of a running animal, the hunter's tracking system is purposely fooled into expecting the animal to arrive at the new location sooner. Such an altered expectation may help compensate for the relatively slow neuromuscular processing.

E. STEREOSCOPIC ILLUSIONS

One of the most elegant examples of visual processing is the way the brain takes the two slightly different retinal images from each eye and creates a three-

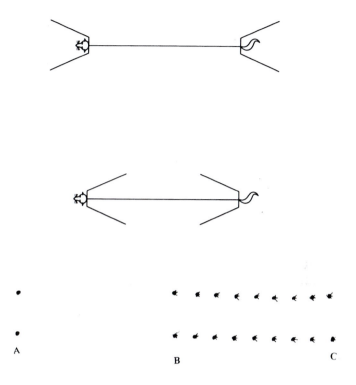

FIGURE 7.22 **(A)** *Top:* A modification of the horizontal Muller-Lyer illusion, showing how the torso of the outstretched animal looks longer than the one with extremities contracted, although both torsos measure the same length! **(B)** *Bottom:* Segment AB = BC. However, the presence of the paw prints makes BC look longer than AB.

dimensional impression (story) in the mind of the beholder. It has traditionally been held that high levels of stereoscopic vision are the reason for the performance of star athletes and pilots. I agree that stereopsis helps make accurate three-dimensional judgments. However, I do not think that stereopsis is the major key to athletic success. Let's look at some basic information.

We have all been taught that we achieve a sense of 3D because we have two eyes. Other depth clues such as shading, object size, and so on, are only of secondary help. The argument is tightly tied in a knot of finality by noting that the monkey skillfully leaps from vine to vine because evolution has placed both its eyes on the front of its face so that it can enjoy stereopsis.

However, there is a very agile branch and high wire negotiator, the squirrel, that displays great agility with its eyes placed on the sides of its head. Then again, the fish eagle can swoop down, capture a fish, and head back into the sky without ever slowing down. Its eyes are also on the side of its head. Admittedly, the eyes of the squirrel and the eagle share some overlap of forward visual field that might be considered a form of stereopsis. However, this type of stereopsis would be very different from animals with both eyes on the front of the face.

A number of years ago, a mother brought her 10-year-old son to see me. He had been struck in one eye by a tree branch. The eye was badly lacerated. A local eye surgeon had repaired it as best he could. What worried the mother the most was the surgeon's prediction of the boy's future. "Since the boy would essentially be one-eyed from then on, he would be limited in his future activities." Because his depth perception would be affected, he should stay away from all "ball" sports. Since he had only one working eye, he should also stay away from any contact activities. Clearly, both mother and boy were upset. I'm not sure how they got my name, perhaps I simply represented a big city specialist, and they wanted another opinion. My feeling was that he should wear protective spectacles when playing a "ball" sport. However, I felt that he should go out for as many ball sports as he wished. As Roger has grown, he has sent me clippings from his local newspaper. He was his high school's leading pitcher and basketball star. Later, he received an athletic scholarship and was able to attend a prestigious college. One-eyed baseball stars are not that rare. The eye doctor for the great home run king, Babe Ruth, wrote a letter to the editor of *Argus,* one of the American Academy of Ophthalmology's publications. In the letter he reported that "The Babe" saw 20/20 out of one eye, but only 20/200* out of his other eye.

Certainly, good stereopsis is important in activities such as flying, where there are few depth clues in the sky. Nevertheless, a friend who had a fine flying record in the U.S. Air Force during the Korean War admitted that he didn't have stereopsis "worth a damn." "How did you manage to pass the depth perception test given all pilots?" I asked. The test depended on pulling a string, which brought a vertical bar to the same location as a reference bar. The observer does the sighting through a small window so as to eliminate all visual clues. "The day before I was to take the test, I watched a colleague with good stereopsis take the test. When both bars were lined up perfectly, I noticed that a mark on the string was perfectly aligned with the edge of the table. When I took the test, I didn't even look at the bars. I simply pulled the string till the mark on the string lined up with the table edge, and passed with the flying colors," he said proudly. Therefore, although stereopsis is useful in making depth judgments, I feel that its role in making these judgments has been overrated.

However, I do believe that there is another extremely important survival role for stereopsis. During the World War II, many important military facilities were painted so as to be hidden by camouflage. For example, simulation of shadows with natural rounded sides was an effective method of hiding airplane hangers.[25] However, pilots with excellent stereopsis were not fooled because the camouflage was flat. Nature too, makes use of camouflage to protect many of its creatures.

---

* In the United States, the definition of legal blindness is a best corrected visual acuity of 20/200 or worse in the better eye.

Figure 7.23 (see color insert) is an example of a snake hidden by blending its skin pattern with the surrounding foliage. In such a situation, only one clue besides the snake's movement will alert the monkey (who is deathly afraid of snakes) or us to the presence of the snake. Stereopsis allows primates to register that the camouflaged snake is sticking out in a third dimension. Once again, subconscious brain processing in the form of stereopsis serves an important survival function. Stereopsis is considered an illusion in this section because it is a new creation, by the brain, of the retinal image, a creation that adds useful meaning.

F. AMBIGUOUS FIGURES

A cloud in the summer sky that looks like the profile of a human face would be an example of the very human trait, where a specific retinal image triggers more than one perception. Since there probably is only a finite number of images in our visual library, it should not be surprising that certain abstract arrangements may trigger human impressions.

In Figure 7.24 we see a well-known variable figure. Is it a picture of a vase or two faces looking at each other? To someone who has never seen a vase, only the

FIGURE 7.24   The famous variable illusion of either a vase or the profiles of two people (exact origin unknown).

profile of the two faces is seen. Hence, the experience of the observer plays a vital role in what is seen. Whereas Figure 24 is almost a generic, variable figure, Figure 7.25 (see color insert) tightly ties the variable options to American culture. Do you see two trees or a profile of George Washington? Every American has been shown this profile of the "father of our country" in elementary school.

Cultural experience as well as immediate context can influence your impression of a figure. Look at the note in Figure 7.26. We are not confused at all by the fact that the same scribble represents an "h" in the word ophthalmologist and a "b" in the word about, or that the "5" in 15% is the same as the "s" in surgery, although the retinal image is obviously the same. There is not a computer today that could read this note correctly. However, our culturally influenced visual brain knows these two words and quickly eliminates any confusion by imposing the proper "story" on each word.

Hence, one might say that life is full of variable figures. Everyone's interpretation is quite unique and private, strongly connected to social, cultural, and family experience, as well as other stimuli being presented at the same time. These influences strongly affect our imagination, and it is our imagination that gives these scenes a reality.

G. APPARENT MOTION

Skilled artists have always been able to generate the perception of motion in a static figure. In fact, there seems to be a special part of the brain that records this apparent motion.[26] Practitioners of "optical art" have created works that appear to move, pulsate, or pop out of the plane of the picture.[27] Ludimar Hermann, a nineteenth-century physiologist[28] produced the grid pattern in Figure 7.27. Notice

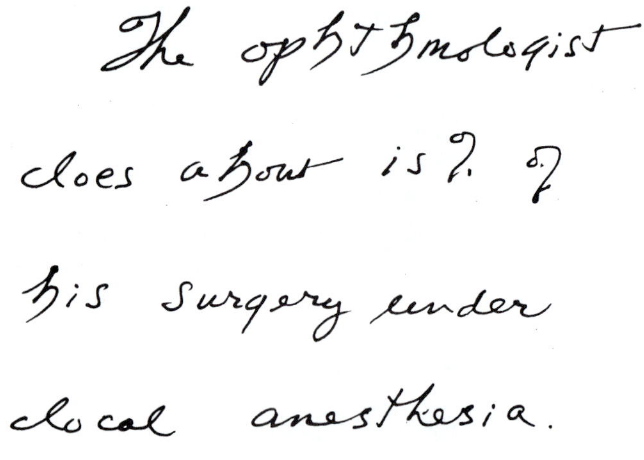

FIGURE 7.26 Written note in which the "h" in ophthalmologist is the same as the "b" in about and the "s" is considered a "5" in the number 15 and an "s" in surgeon.

FIGURE 7.5 A photo of the Mona Lisa by Leonardo De Vinci. Note the local swelling (chalazion) of the left lower lid.

FIGURE 7.7 Versions of a face in which the shadows are given different colors. Only the face with the dark shadows looks real.

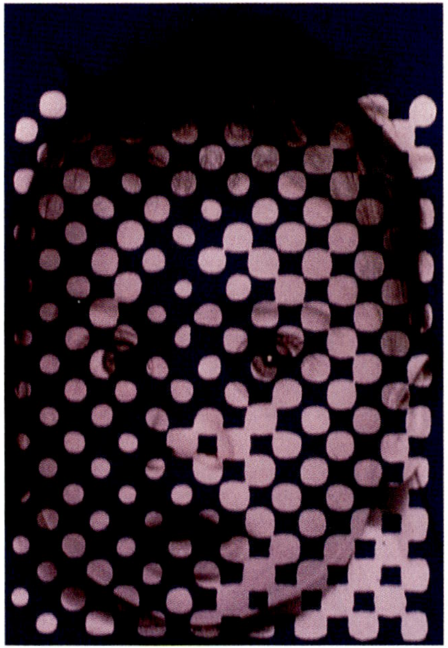

FIGURE 7.12 In this illusion, the vertical bar closest to the trunk of the tree looks longer than the second bar, yet each measures the same height. This illusion remains even when blurred.

FIGURE 7.8 You can still see the face of an infant even though it is buried in medium frequency noise.

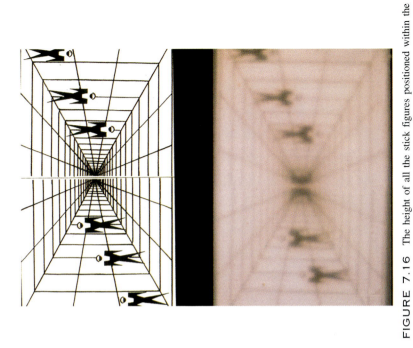

FIGURE 7.16 The height of all the stick figures positioned within the converging lines are all the same size, although the distant figures look considerably larger whether in a fog or upside down. (Modified from Block JR, Yuker HE: *Can You Believe Your Eyes*. Gardner Press, New York, 1989, with permission.)

FIGURE 7.14 A modification of Figure 13, in which the torso of the stick figure with outstretched arms and legs appears larger than the crouched figure, although both torsos are the same length. Once again, the illusion remains, even when blurred.

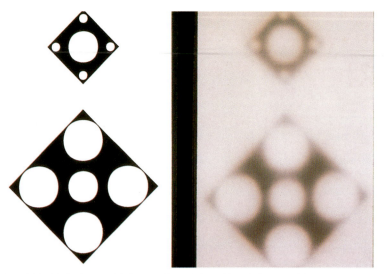

FIGURE 7.17 Both central balls are of equal size, although the central ball surrounded by the small balls looks larger than the one surrounded by large balls. Blurring makes all the balls appear larger, yet the illusion remains.

FIGURE 7.18 Drawing that shows the moon on the horizon to appear larger than the moon in the sky or the reflection in the foreground although all moons are the same size. The illusion seems more exaggerated when blurred.

FIGURE 7.19 The photograph shows natives fishing at sundown. Recall that the angular subtense of the moon and sun are the same (the moon can totally eclipse the sun).

FIGURE 7.23 A Gabon viper hidden within the surrounding leaves. In real life, the three-dimensional aspect of the snake would alert a monkey or human to its presence. (From Marten M, May J, Taylor R: *Weird & Wonderful Wildlife.* Chronicle Books, San Francisco, 1983, p. 159, with permission.)

FIGURE 7.25 A drawing (clear or a blurred version) in which you see either a profile of the figure of George Washington or two trees. (The picture appeared in about 1860, and was called the Tomb and Sham of Washington.)

FIGURE 7.27   The Hermann grid demonstrates pulsating gray blotches at the intersection of the white stripes. Visual physiologists relate this phenomena to the lateral inhibition effects produced by the retinal ganglion cells (From Sekuler R, Blake R: *Perception* (3rd ed.). McGraw Hill, New York, 1994, p. 76, with permission.)

how the gray areas at the intersection of the white stripes appear to pulsate in and out of view.

Figure 7.28 presents a tightly wound spiral that gives the impression of rotating. The rotation simulation seems to depend on a gray triangular area with its apex at the spiral center, fanning out to the periphery and moving around the spi-

FIGURE 7.28   An illusion showing apparent rotation of this spiral. (From Block JR, Yuker HE: *Can You Believe Your Eyes?* Gardner Press, New York, 1989, p. 75, with permission.)

ral. The pulsating gray triangular area is probably related to the mechanism that produces the blotches in the Hermann grid.

Figure 7.29 presents ever-thickening black lines radiating from a central area. The radiating lines are not straight but seem to follow the contour of two concentrically placed tubes or convexities. Note how the more centrally placed circular convexity strongly pulsates (i.e., gray areas between the black lines pulsate). Imagine a coiled snake, gray or tan in color, sunning itself. Overhead, tall sheaves of grass and branches cast dark shadows in a pattern similar to all or a portion of Figure 29. The dark shadows not only seem to accentuate the snake's tubular shape, but the pulsating pattern (Hermann grid phenomenon) calls attention to the convexity. I have used a coiled gray rope (simulated snake), grass and branch simulators, and a simulated sun source (lamp) to create a similar illusion. Therefore, one could speculate that this illusion might be helpful in calling attention to the presence of a snake in shadow. By analyzing the spatial configurations of these depth movement illusions carefully, Byung-Geun Khang was able to maximize the effect in Figure 7.30. Note how the middle column not only stands out in a three-dimensional fashion, but swings sideways when you move your eyes from side to side. Once again, you might imagine this illusion to be used to bring attention to a striped creature hiding in the shadows of the long grass, in which the camouflaging lines do not perfectly line up with the stripes of the creature.

FIGURE 7.29 Modification of a radiating pattern in which the dark lines over the central round convexity strongly pulsate. Such an illusion under the right conditions might call attention to a coiled snake, covered by linear shadows created by tall grass or branches. (The original pattern is from Wade N: *Psychologists in Word and Image*. MIT Press, Cambridge, MA, 1995, p. 162, with permission.)

FIGURE 7.30 An illusion created to accentuate the middle column, which not only pops out but swings sideways when your eyes move back and forth across the column. (From Khang B, Essock EA: Motion illusion from 2-D periodic patterns. *Invest Ophthalmol. Vis. Sci.* Feb. 16, 1996, 37(3):5743, with permission.)

Finally, mention should be made of the phenomenon of the flashing lights on a cinema marquee that seem to show a light moving around the border of the marquee.[39] A similar effect results when we watch a movie composed of rapidly changing frames. Connecting the frames to give the perception of smooth movement (apparent movement) would be quite useful if one were watching a running animal behind a series of trees. The smooth movement perceived gives the separate images a useful story.

H. DISAPPEARING IMAGES

There are a number of situations when the eye/brain axis makes an image disappear. For example, during an eye movement, conscious vision does not take place. In fact we really do not see accurately for about 100 msec prior to a saccadic eye movement, during the movement, and for about 50 msec after the close of such a movement. This phenomenon is dependent on a feed-forward mechanism that announces that a saccade is about to take place; thus the visual process is turned off.[29–31]

A similar occurrence takes place when we move our head, body, or eyes. Clearly, images from the outside world race across our retina yet we do not get the impression that the world is moving. Somehow the act of movement or the signal to start movement is able to command the visual brain to cancel the impression of movement. Both of the phenomena described optimize the act of vision.

Another disappearing act is known as the *stabilized image*.[32] Using an unusual contact lens and projection system, an image was formed on a very specific part of the retina, when no amount of eye quiver (microsaccade) could move the image to another retinal location. Under these circumstances the image disappeared. This same disappearance of the retinal image took place if the eye muscles were paralyzed with curare[33] and an image was projected onto one specific retinal area. However, in this case a combination of physiologic events occur. First, such disappearance represents a kind of burnout of the retinal receptors and such burnout is prevented by the normal pattern of eye movement during breathing, microsaccades, or natural blinking. Second, since certain types of images disappear faster than others, it is felt that the mechanism is modulated by the visual brain.

I. GLOBAL IMPRESSIONS COME FIRST

I can recall a bright little boy in the 6th grade who could recall the dates of all the great battles, but did not know which side won the war. He had lost the forest through the trees. Our visual system does not work that way. In fact, it may see the forest and very little of the trees. In Figure 7.31 is the "Santa down the chimney illusion." Notice that the grouting (fill between the bricks) is darker on two sides. This makes us think that the bricks, themselves, are also darker. Only care-

FIGURE 7.31   The Santa up the chimney illusion. Notice that because the grouting between the bricks is darker on one side, the bricks themselves appear darker, but really are not. (From Livingstone M, Hubel D: Through the eyes of monkey and man. In *The Artful Eye* (Gregory R, et al, eds). Oxford University Press, Oxford, 1995, pp. 52–65, with permission.)

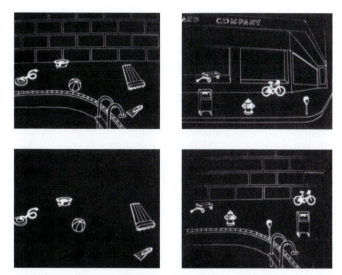

FIGURE 7.32   The objects in the lower panel are difficult to identify. However, within the context of a pool setting they are easy to identify. (From Boyce JJ, et al: Effect of background information on object identification. *J. Exp. Psych. Human Percep.* 15(3):556, 1989, with permission.)

ful observation shows the bricks to be of uniform color throughout. The woodcut of Shakespeare (Figure 4) is another example of why most of us never really noticed that the picture shows Shakespeare with two right eyes. This big picture approach of the visual system also fills in the blind spot (which we all have), so that we do not notice the normal blind area within our field of vision.

Finally, this global approach can help us to identify certain ambiguous objects by providing a relevant context. The top right diagram in Figure 7.32 presents a number of objects in a pool setting. Notice how difficult it would be to identify the same objects if isolated (lower right).

In a dangerous situation, where one must act quickly, I can appreciate the virtues of a visual system that can quickly present a global view. On the other hand, it is nice to know that one can be trained to notice details within a scene.

SUMMARY

A. SUBJECTIVE ASPECTS OF VISUAL ILLUSION

Visual illusions might be looked on as somewhat evolutionarily derived reflexes that helped our primitive ancestors survive. However, they are not quite inherited reflexes, but inherited inclinations toward reflex activity, which can be hard wired during childhood if the culture considers them important. Professor V.S. Ramachandran of the University of California has been one of a few scientists who support the idea. To him, "illusions represent a series of tricks that the evolutionary process has bumped into and retained to enhance survival."[34]

The role of an individual culture in enhancing or negating certain illusions deserves a bit of amplification. The pygmy living in a dense forest has no need for the size constancy needed on an open plain. People living along the Amazon would not see a fountain or George Washington in the variable figures presented in this chapter. On the other hand, all cultures inherit the ability to quickly recognize friendly and alien faces. Such recognition is dependent on many aspects of experience. When I presented the illusion of a crowd of faces at a baseball game, only the older doctors quickly recognized President John Kennedy in the crowd. The automatic identification of friend or foe may have also been important in ancient times. We automatically suspect someone who looks different. In older times, such a "reflex" may have helped identify someone from an alien village. In truth, we see the way our culture teaches us to see.

This idea of sincerely seeing and reporting everything in the context of one's local tribal values, rather than in absolute terms, might seem to be an alien idea to "modern, rational humans." We moderns prefer to feel that the registration of reality can be objective and not contaminated by social bias or inherited neurologic circuitry. In the sciences, we like absolutes. The modern human is comfortable discussing absolute ceilings in aviation, absolute pitch in music, absolute zero temperature in physics, and absolutes in philosophy. We feel secure with a solid number. Look at the program for a football game. The height and weight of each player is listed precisely. The listing on a container of ice cream clearly notes the calories, percentages of each component, cholesterol level, and so on. We like the idea of objective reporting and objective referees. Our love affair with numbers and the concept of hard data has led us to believe that there can be objective reports, objective referees, and objective judges. Yet, in this chapter we have met Dr. Sacks' patient Dr. P., who was an objective reporter, but had lost his humanness. In normal humans, the image projected on the retina is usually not the same as the object we perceive. I do not view this lack of objectivity in our visual system to be a deficit, but rather a sign of our humanness. The quote in Professor Zajonc's book,[35] "There is no such thing as a ball or a strike till the umpire calls it," seems to crystallize this idea. Indeed, all of our visual brains differ, one from another, and it is the visual brain that is the umpire.

B. GLOBAL IMPRESSION VERSUS DETAIL DETECTION

Recall how Sherlock Holmes' feats of observation made him quite different from Dr. Watson. Holmes had trained himself to erase away the global impression of a scene and focus on minute details, such as the tobacco stain on a finger or the type of dirt caught along the seam of a shoe. He also had learned to notice the importance of the absence of an expected detail, such as the absence of barking by a normally vigilant watch dog. It is very human to get the general impression of a scene and not notice details, as was Dr. Watson's style. However, it is equally human to be able to go through training so as to become expert at noticing details (the hallmark of Sherlock Holmes).

In a sense, children of the twentieth century are inadvertently trained to examine visual details as in no century before. Aside from school work and picture puzzles, exploded views appear on cereal boxes as well as instructions for assembling toys. The ubiquitous computer forces all of us to align the mouse icon in the precise manner.

A study examining the drawings of faces as well as block designs by nine different groups from hunter gatherers to literates gives further support to these ideas. The results of the study suggested that recognizing intrapattern details, in contrast to differences between objects, emerges with literacy but is not yet actualized in preliterates (whose survival requires quick flight or fight response upon prompt assessment of danger).[38]

C. FACIAL RECOGNITION

This chapter has attempted to demonstrate the power of our facial recognition system. The following anecdote demonstrates that the system is housed in a specific region of the brain.

Soon after World War II, the German psychiatrist Joachin Bodamer treated a young man who had suffered a bullet wound to a specific portion of the brain. The young man complained of a very unusual problem. Since the injury, he could not recognize his own face or that of any of his friends or family. When shown a picture of a dog, he said that it looked like a person with "curious hair." Bodamer called the condition prosopagnosia (from the Greek *agosia* meaning no knowledge of, and *prosopon* meaning face).[36]

Recently, two English psychologists, Vicki Bruce and Andrew Young[37] have introduced a theory that describes the steps that the brain might take to process the presentation of a new face. After the retinal image is sent for processing, the brain simply registers the physical appearance of the face. This information is passed on to the elements of level two, which registers whether the face is familiar or not. Data concerning the familiar face are passed on to level three, where biographic information stored within the brain is gathered, giving the face a richer identity. The person's name, however, is stored in another portion of the brain, independent of the personal information. Such a separation would explain the way some of us might identify everything about a person, yet not recall their name. In a patient with prosopagnosia, aging, disease, or injury might effect any or all of the processing levels.[13]

Note that the monkey also has a facial recognition center.[38] The social network of the monkey is a key element in their survival. Indeed, this was expressed succinctly by the investigator Watson, "A solitary chimpanzee is no chimpanzee."[35] However, the fact that adults can recognize blurred faces, children can recognize upside down faces, and some newborn babies can imitate facial expressions on their first day of life, all illustrate that humans have developed an even higher

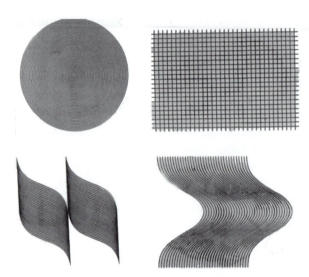

FIGURE 7.33 Can you see the faces subtly introduced in four different geometric patterns? (From Wade N: *Psychologists in Word and Image*. MIT Press, Cambridge, MA, 1995, with permission.)

level of this function. Our emphasis on facial recognition is so strong that we see faces when they are only subtly presented. In Figure 7.33 Nicholas Wade has faintly added facial features to four geometric patterns. How quickly do the faces pop out for you? This strong emphasis on the recognition of faces and facial expression points to a very important human trait.

REFERENCES

1. Marr D: *A Computational Investigation ino the Human Representation and Processing of Visual Information*. W.F. Freeman, San Francisco, 1982.
2. Cramer F: *Chaos and Order*. VCH Publishers, New York, 1993.
3. Southall JPC: *Mirrors, Prisms and Lenses,* MacMillan, New York, 1933, p. 592.
4. Burton M: *The Sixth Sense of Animals*. Taplinger Publishing, New York, 1972, pp. 118–132.
5. Stratton GM: Some preliminary experiments in vision without inversion of the retinal image. *Psych. Rev.* 3:611–617, 1886.
6. Horton JC, Hoyt WF: The representation of the visual field in human striate cortex. *Arch. Ophthalmol.* 109:816, 1991.
7. Mountcastle VB: The view from within: Pathways to the study of perception. *John Hopkins Med. J.* 136:109–131, 1975.
8. Gombrich A: *Art and Illusion,* Princeton University Press, Princeton, NJ, 1956.
9. Sacks O: *The Man Who Mistook His Wife for a Hat*. Harper Perennial, New York, 1990.
10. Lettvin J, Maturana B, McCulloch C, Pitts R: What the frog eye tells the frog brain. *Proc. Inst. Radio Eng.* 47:1940–1951, 1959
11. Goldstein K, Gelb A: Psychologische Analysen hirnpathologischer Falle auf Grund von Untersuchunger Hirnverletzer. I Abhandlung: Zur Psychologie des optischen Wahrnehmungs-und Erkennungsvorganges. *Z. Ges. Neurol. Psychiat.* 41:1–142, 1918.

12. (Quoted in) Demott D: *Teaching What We Do.* Amherst College Press, Amherst, MA, 1991.
13. Piaget J: *The Mechanism of Perception.* Basic Books, New York, 1969.
14. Rock I: *An Introduction to Perception.* Macmillan, New York, 1975.
15. Peli E, Goldstein K, Young GM, Trempe CL: The critical spatial frequency for face recognition. *Digest of the Topical Meeting on Noninvasive Assessment of the Visual System.* Technical Digest Series, Vol. 1, Santa Fe, Feb. 4–7, 1991, Optical Society of America, pp. 105–108.
16. Benson PJ, Perrett DI: Perception and recognition of photograph quality caricatures: Implications for the recognition of natural images. *Eur. J. Cog. Psych.* 3:105–135, 1991.
17. Grüsser OJ, Selke T, Zynda B: A developmental study of face recognition in children and adolescents. *Hum. Neurobiol.* 4:33–39, 1985.
18. Berndl K, Grusser OJ, Martin M, Remschmidt H: Comparative studies on recognition of faces, mimics and gestures in adolescent and middle-ages schizophrenic patients. *Eur. Ach. Psychiatr. Neuro. Sci.* 236:123–130, 1986.
19. Segall MH, Campbell DT, Herskovits MJ: *The Influence of a Culture on Visual Perception.* Bobbs Merrill, Indianapolis, 1966.
20. Lee ND: The optic flow field. The foundation of vision. *Phil. Trans. R. Soc. London B* 290:169–179, 1980.
21. Turnbull CM: *The Forest People.* Simon and Schuster, New York, pp. 247–253, 1961.
22. Block JL, Yuker HE: *Can You Believe Your Eyes?* Gardner Press, New York, 1989.
23. Sarton G: Lunar influences on living things. *Isis* 30:501, 1939.
24. Stein H, Slatt B: *Hitting Blind.* Don Mills, Ontario, Musson, 1981.
25. Lukiesh B: *Visual Illusions.* Dover, New York, 1965.
26. Zeki S, Lamb M: The neurology of kinetic art. *Brain* 117:607, 1994.
27. Oguchi H: *Japanese and Geometric Art.* Dover, New York, 1977.
28. Sekuler R, Blake R: *Perception* (3rd ed.). McGraw-Hill, New York, 1994, p.76.
29. Miller JM, Buckisch J: Where are the images we see. *Nature* 386:549, 1997.
30. Schlag J, Schlag-Rey M: Illusory localization of stimuli flashed in the dark before saccades. *Vis. Res.* 35:2347, 1995.
31. Volkman EC, Schick ML, Riggs LA: Time course of visual inhibition during voluntary saccades. *J. Opt. Soc. Am.* 58:562, 1969.
32. Prichard RM: Stabilized images on the retina. *Scientific American* 204:72, 1961.
33. Stevens JK, Emerson RC, Gerstein GL, Kallos T, Neufield GR, Nichols CW, Rosenquist AC: Paralysis of the awake human: Visual perception. *Vis. Res.* 16:93, 1976.
34. Ramachandran VS: How the brain adds meaning to movement detection. Guest Editorial. *Perception* 14:97–103, 1985.
35. Zajonc A: *Catching the Light.* Bantam Books, New York, 1993.
36. Grüsser O, Landis T: *Visual Agnosias and Other disturbances of Visual Perception and Recognition.* CRC Press, Boca Raton, FL, 1991, p. 259.
37. Szpir M: Accustomed to your face. *American Scientist* 80(6):537, 1992.
38. Pontius AA: Spatial representation in face drawings and block design by nine groups from hunter-gatherers to literates. *Percept. Motor Skills* 85 (3, part):947–959, 1997.
39. Johansson G: Visual motion perception. *Scientific American* 232:76–88, 1975.

8

BRAIN SHARPENING OF THE RETINAL IMAGE

1. Brain Mechanisms that Enhance the Retinal Image
   A. *Filling in Information*
   B. *Contrast Enhancement*
   C. *Edge Sharpening*
   D. *Vernier Acuity*
   E. *Removing Distractions*
2. Vision Function and Aging

1. BRAIN MECHANISMS THAT ENHANCE THE RETINAL IMAGE

There is a group of visual phenomena in which the retinal image is enhanced or made complete by the brain. They represent ways of improving the retinal image in nonoptical ways. One might think of these brain processing effects as an example of a method of going beyond the limits of the laws of optics to bring out visual information.

A. FILLING IN INFORMATION

Have you ever watched a child and a father play peek-a-boo around a tree, or a mother play peek-a-boo from behind a book with her infant? The object (the tree or the book) (Figure 8.1) covers most of the parent, yet the child knows the parent is whole and present. The child's giggle, once the parent is completely revealed, probably represents relief that the parent did not really disappear. As the child gets older, the game of peek-a-boo loses its luster because the child now understands that the parent will always be there.

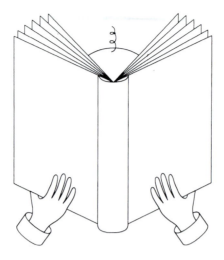

FIGURE 8.1   The appearance of the top of a head behind a book suggesting the face is there as well.

Figure 8.2 is a more adult version of the above illusion. Because of our previous visual experience, we assume that the partially covered words represent "THE EYE". Wrong!

If there is a discontinuity in an object, we almost always assume that the object is partially covered. We simply create a full impression of the covered object in our mind. In Figure 8.3, we are made to think that one geographic

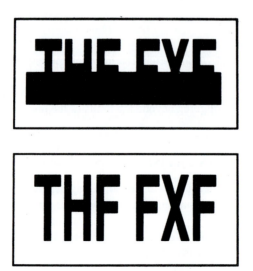

FIGURE 8.2   Because of our previous visual experience, we assume the covered words will spell "THE EYE". Upon removal of the cover, we see we were wrong.

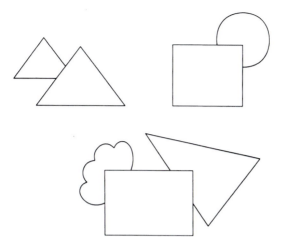

FIGURE 8.3   A series of geometric shapes with some appearing to cover the others. In fact, the edges of the shapes simply abut each other.

form covers the other, although the artist simply fitted the outline of one object next to the other.

These illusions illustrate the ability of our visual processing system to fill in all sorts of blanks and incongruities within the retinal image (including the physiologic blind spot) in order to create a coherent story. With so many dangerous or important objects in our world partially covered, wouldn't you think that this storytelling ability of the visual brain is important for human survival?

Let us suppose that the object of regard is not covered, but the observer has a brain lesion that has produced a small scotoma in the visual field. When such a patient is presented with a circle or square, so that part of the figure resides inside the scotoma, the patient, after a brief period,[1,2] will report the gap has filled in and the figure looks whole. The same phenomena takes place if part of an image falls on the physiologic blind spot. It suggests that the visual system, faced with a gap in the information hypothesizes (gambles) that the region surrounding the scotoma has the needed data and places that data within the scotoma to produce a complete scene.

There is another series of illusions, known as gap figures, which should be included in this section. Figure 8.4 shows an example of a well-known gap illusion created by Kanisza.[3] Although part of the pattern is absent (gap), the gap takes on an identity (a triangle), as opposed to ignoring the gap and simply filling in the absent segments of the round balls. The defect (gap) is strongly highlighted much as a fresh footprint might appear highlighted to a skilled guide. In fact, natural phenomena like the footprint might be the very reason for the "Kanisza" phenomenon.

As noted in the previous chapter, in Conan Doyle's story "The Adventure of Silver Blaze," the detective Sherlock Holmes identifies the murderer when he learns that the watch dog must have known the murderer (since no barking was

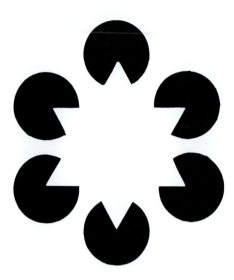

FIGURE 8.4 A gap figure (after Kanisza), in which a solid triangle seems to cover the gaps in the dark disks. (From Kanisza G: Subjective contours. *Scientific American.* 234:48–52, 1976, with permission.)

heard on the night of the murder). As in the Kanisza figure, the absence (gap) in the expected pattern of the dog's behavior takes on a heightened importance. Hearst[4] appropriately called this phenomenon, "getting something for nothing."

Finally, mention should be made of our failure to notice the fleeting disappearance of an image during a blink, a twitch, a flicker, or a saccade. Since all these intervals last for less than a second, the probability is high that no life-threatening event will take place during these short periods of blackout.

B. CONTRAST ENHANCEMENT

When the eye clinician measures visual ability, he or she may ask the patient to read a chart of black letters against a white background. Each line of letters is smaller than the one above. A threshold is reached when the patient comes to a set of letters that is simply too small to resolve. Throughout the test the patient is never asked about the contrast of the letters against the background. Figure 8.5 presents a series of circles (Landolt rings) containing a black broken ring against a background of ever-increasing grayness. As the gray background gets darker, the contrast obviously gets lower; in fact, the measurement of contrast sensitivity is probably the most important indicator of how well we see objects in the real world. For example, the skin of many animals is often shaded (gets lighter from back to abdomen) so that their contrast is lowered against their natural background, and they appear "invisible." In Figure 8.6 (see color insert), we see a penguin leaping into the water. To a predator below, the penguin's

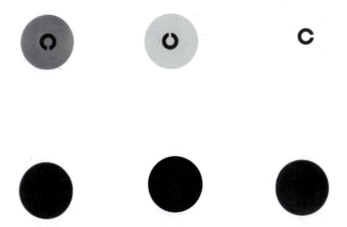

FIGURE 8.5   A series of dark circles with small gaps (Landolt rings) seen against darker and darker backgrounds in which the contrast gets lower.

white belly produces a poor contrast against the sky above. To a potential predator above, the black back provides poor contrast against the dark surrounding sea.

Interestingly, the visual brain has the ability to sharpen the contrast of elements in the retinal image. Figure 8.7 presents two faces of the same grayness. In

FIGURE 8.7   *Right side:* Gray face is seen against a black background. *Left side:* Same gray face is seen against a white background. A contrast enhancement function makes the gray face look lighter on the black and darker on the white side. Bottom: (Illusion created by R. Miller, F. Miller, and D. Miller.)

the right figure, the face is seen against a black background, while in the left figure against a white background. Can you appreciate what the brain has done? The gray face on the black background looks lighter (enhancing its contrast), while the gray face looks darker on the white background. This effect is reduced when the edges of the faces are fuzzy. One might speculate that a sharper edge on an object brings out a stronger contrast enhancement response.

C. EDGE SHARPENING

If you wear glasses for myopia or hyperopia, take them off for a moment and look across the room at a framed picture. True, you cannot see any of the details within the picture. However, notice the edges of the frame. There is a distinct boundary against the wall. The visual system places a priority on sharpening the edge of the retinal image, even though the details within remain hazy. One might suppose that in the case of the picture on the wall, recognizing the edges of the frame at least tells you that the fuzzy blob on the wall is a picture rather than a gaggle of insects.

There is a second way that the brain works on the edge of a retinal image. If two objects of similar brightness are placed next to each other, they should appear to merge into one object. However, if the connecting edge of one of these similar objects is a bit darker than the connecting edge of the other, then the entire side with the darker edge appears darker than the lighter one. The greater contrast at the edge has spread across the whole panel.

This phenomenon, known as the Craik-Cornsweet-O'Brien illusion, is seen in Figure 8.8 (see color insert). If you put your finger or a pencil between the two vertical rectangles (occluding the boundary), you will see that they are actually the same brightness.

Could there be a survival value in such a process for our early hominid ancestors? Admittedly, this example is a bit of a stretch of the imagination, but this phenomenon does enhance the appearance of animal footprints in the mud, dirt, or grass. At the edge of the footprint, there is a depression on the inside and a heaping up on the outside. With the sun in the proper position, the heaped-up areas could reflect light to one side, creating an adjacent shadow on the other side. The Craik-Cornsweet-O'Brien mechanism increases the contrast of the whole footprint and makes it easier to detect. Admittedly, we moderns do not commonly make use of this illusion. However, it is possible that hunting societies may have benefited from it.

Figure 8.9 (see color insert), represents an example of how the visual system works with the other sensory systems to identify shapes. In the illustration, a frosted glass makes the rose and apple appear as two red blobs. When the visual system cannot separate the two, then our sense of smell or use of prior history often helps make the differentiation (i.e., the equivalent of improving the contrast and resolution).

FIGURE 8.6   Photograph of a penguin leaping into the sea. Its white belly provides a poor contrast against the sky, while its dark back provides a low contrast against the dark sea.

FIGURE 8.8   The Craik-Cornsweet-O'Brien illusion. The perceived difference in the darkness of the vertical panels disappears if you place a pencil or your finger along the border between the panels. The effect persists even if the panel details are blurred.

FIGURE 8.9 The observer sees two fuzzy red blobs through the frosted glass. Aside from visual clues, we use our sense of smell or knowledge of prior history to distinguish the two. (From Cheney M: Inverse Boundary Value Problems. *American Scientist* 85:452, 1997, with permission.)

FIGURE 8.11 Note the deer hidden in the bush. By almost fully closing your eyes, the distraction of the high grass is cancelled and the animal is easier to see. (From Osborne C (Ed.), Tanner O (author): *Animal Defenses,* Time-Life Films, New York, 1978, with permission.)

D. VERNIER ACUITY

In an earlier chapter it was noted that a normal sighted human being (20/20 visual acuity) could detect a separation between two objects as small as one minute of angular subtense. Interestingly, it was said that the great outfielder of the Boston Red Sox, Ted Williams, had a visual acuity of 20/10 (could detect a $1/2$ minute of angular separation). However, there is a visual task (Vernier acuity), which has a threshold of about 5 seconds of arc ($1/12$ of a minute of arc). Indeed, most normal sighted people can line up dots or notice a discrepancy in the alignment of dots or lines, as small as 5 seconds of arc or less. Figure 8.10 demonstrates two typical alignment tasks presented to experimental subjects. We know that the brain is involved in the processing because the experiment can be redesigned so that one eye is presented with the top and bottom dot (which are in alignment), while the other eye is allowed to see only the middle dot. Subjects show similar thresholds when this binocular experiment is done. Clearly, the brain must be the location of the ultimate processing in these experiments. Since a threshold of less than 5 seconds is well beyond the optical diffraction limit of the eye, vernier acuity represents a very special example of sophisticated brain processing. The challenge, once again, is to try to conceptualize some important task, necessary for the survival of our homo sapien ancestors, which required such precision. Incidentally, we also know that the macaque monkey demonstrates a high level of vernier acuity. In fact, the adult monkey averages a threshold of about 13 seconds of arc.[5] Thus, it is possible that monkeys and early humans used vernier acuity to detect the presence of an animal or an enemy hid-

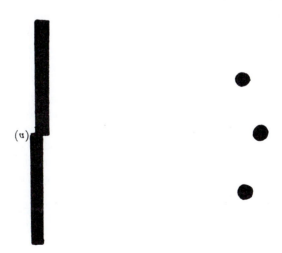

FIGURE 8.10   Two typical targets for measuring vernier acuity. The subject either notes misalignment of the vertical lines or of the three dots.

ing behind a stalk or a tree, by noting a misalignment between the edge of the tree or stalk and the protruding body of the enemy. If that possible scenario was true, then normal vernier acuity might save a life. Once again, brain enhancement of the retinal image brings out details well beyond the limits of the best optical resolution.

E. REMOVING DISTRACTIONS

In Figure 8.11 (see a color insert), a deer is hidden in the high grass. By closing your eyes almost completely you create a blurring of the overlying fine grass, and the animal is more vividly seen. A similar effect can be achieved through a heavy rain. By almost closing your eyes, figures will be seen more easily through the rain. These are examples of erasing the distractions that have a high spatial frequency. In Figure 7.8 in the last chapter, one can easily recognize the face of a baby behind a web of distracters (i.e., wire mesh). A normal subject can also follow a story on the T.V. screen in the face of snow and static. Interestingly, patients with certain visual cortex lesions cannot follow a T.V. presentation if there is a static overlay.

2. VISUAL FUNCTION AND AGING

One might describe our visual system as one so sophisticated that it literally thinks for itself. It might be fair to ask if the system loses any of these sophisticated features as the human ages.[6–11] Recall that in a previous chapter on the elements that produce the retinal image, it was noted that the eyes of most of us function well into our 70s and 80s. On the other hand, the visual processing area of the brain seems to lose some if its "high tech" functions with age. Devaney and Johnson have examined, at autopsy, the visual brain in 23 human subjects aged 20 to 87 years.[12] Their work shows a slow but steady decline in the neuron density with age.*

Figure 8.12 is a graphic presentation of their research. If we consider the density of nerve cells to be a maximum 100% for the 20-year-old, we note that at age 40, the neuronal density has decreased by about 15%. At age 60, the density has dropped by about 30% from the level of the 20-year-old. The 80-year-old winds up with about 46% of the density of the 20-year-old. Remember that these represent average figures, which means there will be some among us that lose far fewer neurons and some who will lose more.

From a practical point of view, what does this loss of nerve cells mean? First, our visual reaction time becomes slower. The older person cannot interpret a

---

* Neuron population densities in macular projection area of 23 human subjects, 20 to 87 years of age. Neuron density decreased 54% from approximately 46 million neurons per gram of tissue at age 20, to approximately 24 million neurons per gram of tissue by 80 years.

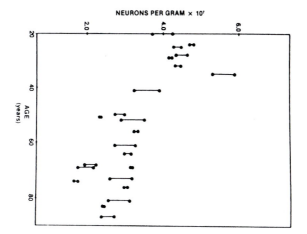

FIGURE 8.12 A graphic display of age versus neuron density of the macular projection area of the visual cortex. The data are from 23 human autopsy specimens, ages 20 to 87 years of age. Note how the number of nerve cells in the brain drops by about 50% from age 20 to 80. (From Devaney KO, Johnson HA: Neuron loss in the aging visual cortex of man. *J. Gerontol.* 35:836, 1980, with permission.)

road sign if he or she drives by it too quickly. At a party or convention, it becomes more difficult reading the name tag of someone moving through the crowd. In fact, in a study comparing 20-year-olds with 64-year-olds, David Walsh of the University of Southern California found that the older group took twice as much time to process visual information.[13] Interestingly, a brighter target reduced the visual processing time in subjects of all ages.[14] That may be one of the many reasons why older folks see newspaper print better in good light.* However, the most striking application of these findings relates to vehicle crashes in older drivers. As a background, let me point out that in all states in the United States, a driving license is not issued unless the driver can read between 20/40 and 20/60 (depending on the individual state law) with his or her better eye. However, there has never been a scientific study that could show a relationship between poorer visual acuity (the retinal image) and an increased frequency of car crashes. Then Dr. Karen Ball and her colleagues[15] devised a clinical test to check the pertinent vision functions related to driving.† Their results show a strong correlation between auto accidents and the test scores of

---

* A brighter reading light also helps patients with age-related macular degeneration and/or cataracts see reading material better.

† The test had three parts. It measured visual processing time, ability to divide attention between central and peripheral targets, and ability to notice a peripheral target embedded in distraction or noise.

visual functions handled by the brain. Thus, those older subjects who have been involved in more than their expected share of auto accidents not only reacted more slowly, but were more distracted by visual noise, and failed to notice events with their peripheral vision, when concentrating on the center of the road. Therefore, this group of elders were poorly equipped to process larger amounts of visual data or cancel visual noise. These results explain how an elderly patient with healthy eyes and 20/20 visual acuity may legitimately complain of reduced vision.* Unhappily, we have no simple clinical tests to measure these functions. This suggests that the vision processing section of the brain may wear out before the elements of the eyes.[16]

This phenomenon of wearing out at an earlier age suggests that certain elements of the vision system might be a relatively new product of the evolutionary process, and thus less resilient to the aging process (decrease in speed of processing information, easier distractibility, and diminished focus). Of course, this is all mere speculation.

REFERENCES

1. Lee, DN: The optic flow field: The foundation of vision. *Phil. Trans. R. Soc. London B.* 290:169–179, 1980.
2. Ramachandran VS: 2-D or not 2-D, that is the question. In *The Artful Eye* (Gregory R et al, eds.). Oxford University Press, Oxford, 1995, p.250.
3. Kanisza G: Subjective contours. *Scientific American* 234, No. 4, 1976.
4. Hearst E: Psychology of nothing. *American Scientist* 79:432, 1991.
5. Tang C, Kiorpes L, Morshon JA: Stereo acuity and vernier acuity in Macaque monkies. *Invest. Ophthalmol.* Abstract Book 36, No. 4, 5365, March 15, 1995.
6. Beck EC, Dustman RE, Shenkenberg T: Life span changes in the electrical activity of the human brain as reflected in the cerebral evoked responses. In *Neurobiology of Aging* (Ordy JM, Brizzee KR, eds.). Plenum Press, New York, pp.175–192.
7. Farkas MS, Hoyer WJ: Processing consequences of perceptual grouping in selective attention. *J. Gerontol.* 35:207–216, 1980.
8. Ordy JM, Brizzee KR, Hansche J: Visual acuity and foveal cone density in the retina of the aged rhesus monkey. *Neurobiol. Aging* 1:133–140, 1980.
9. Rabbit PMA: An age decrement in the ability to ignore relevant information. *J. Gerontol.* 20:233–238, 1965.
10. Sekuler R, Ball K: Visual localization: Age and practice. *J. Opt. Soc. Am. A.* 3:864–867, 1986.
11. Wright LL, Elias JW: Age differences in the effects of perceptual noise. *J. Gerontol.* 34:704–708, 1979.
12. Devaney KO, Johnson HA: Neuron loss in the aging visual cortex in man. *J. Gerontol.* 35:836–841, 1980.
13. Walsh D: The development of visual information processes in adulthood and old age. In *Aging and Human Visual Functions* (Sekuler R, Kline D, Dismukes K, eds.). Alan R. Liss, New York, 1982, pp. 30–55.

* In fact, the elderly population in the United States has (overall) good visual acuity. In a study involving over 2,500 such subjects, about 93% could see 20/40 or better, with both eyes open, suggesting that at least one eye had an acuity of 20/40.[5,8]

14. Plude D, Hoyer W: Attention and performance identifying and localizing age deficits. In *Aging and Human Performance* (Charness N, ed.). Wiley, New York, 1988, pp.47–99.

15. Ball K, Owsley C, Sloane ME, et al: Visual attention problems as a predictor of vehicle crashes in older drivers. *Invest. Ophthalmol. Vis. Sci.* 43(11):3110, 1993.

16. West S, Rubin GS, Muũoz B, Schein OD, Fried LP: Function and visual impairment in a population based study of older adults. *Invest. Ophthalmol. Vis. Sci.* 38–70, 1997.

9

COLORING THE RETINAL IMAGE

Introduction
1. Neural Processing of Color
2. Increased Brightness Under Yellow Light
3. Color Constancy
4. Enhancement of Color Contrast
5. Color Blindness
6. Evolution of Color Vision
7. Other Advantages of Brain Processing of Color Vision

INTRODUCTION

There must be something both challenging and titillating about color vision, because its study has attracted some of the most brilliant scientific minds of the Western world ("Giants of Light"). One can only speculate as to the reason. Newton may have identified its essential intrigue when he observed "that there is no color in the physical world. The sensation of color is created inside us." Who was Sir Isaac Newton? Recall that his experiments* showed that light travels in straight lines, and that white light can be broken into colors by a prism. He theorized that light must be made up of a stream of submicroscopic particles that constitute rays.[22] Figure 9.1 shows a drawing by Newton that depicted his conception of what the prism did to the light. Parenthetically, many scientists often describe their work as fun (i.e., a form of play). Although play is not strictly a visual endeavor, I think that combining play and work may also be a very human trait.

Newton himself wrote, "I do not know what I may appear to the world, but to myself I seem to have been only like a boy, playing on the seashore, and diverting

---

* When Cambridge University was closed due to the outbreak of the plague, Newton went home, where he performed these experiments.

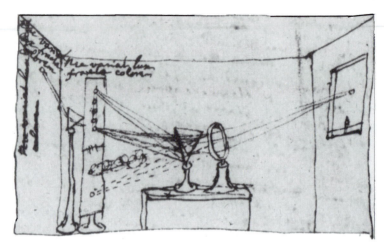

FIGURE 9.1   Copy of Newton's original diagram depicting the effect of a prism on a beam of white light (From Wertenbaker L, Editors of U.S. News Books: *The Eye*. U.S. News Books, Washington, D.C., 1981, p. 20, with permission.)

myself in finding a smoother pebble or a prettier shell than ordinary, while the great ocean of truth lay undiscovered before me." Newton's description of himself suggests three qualities that all the "Giants of Light" possessed. As Gary Zukav noted in *The Dancing Wu Li Masters,*[1] "the first is a childlike ability to see the world as it is and not as it appears according to what we know about it (a culturally induced illusion). The second characteristic may also be called childlike. It is the ability to see as obvious that which to the rest of the world sees as nonsense. Finally, it is also a childlike innocence which knows not of the reprisals of the established order, when an accepted concept is challenged."

Newton must have appeared quite radical when he wrote, "For the rays, to speak properly, are not colored. In them there is nothing else than certain power and disposition to stir up a sensation of this or that color."

But how did the human eye actually "stir up the sensation of color?" Was there a special retinal receptor for each color? The answer to this question was first proposed by another brilliant young graduate of Cambridge University. In 1801 Thomas Young wrote, "Now as it is almost impossible to conceive each sensitive point of the retina to contain an infinite number of particles, each capable of vibrating in perfect unison with every possible undulation. It becomes necessary to suppose the number limited, for instance, to the principle colors, red, *yellow,* and blue." Parenthetically, he changed the composition of the three primary colors to red, *green,* and violet soon after.* It might be worth a moment to find out what sort of a person was Thomas Young.

---

* Recent microspectrophotometry experiments have shown that the human retina has three different types of cone receptors. One has a peak sensitivity to blue light, another a peak sensitivity to green light, and the third to red.

Newton's idea that light was made of particles that moved in straight lines dominated optics for over 100 years. Newton's ideas prevailed until they ran into those of Thomas Young. Always ahead of his teachers, he taught himself fourteen languages, the classics, physics, history, instrument making, and bookbinding, to name but a few of his accomplishments. It was no surprise that his fellow medical students at Cambridge called him "Phenomenon Young." Helmholtz later called him "the last man who knew everything."[2] At age 26, after graduation from medical school, Young read a bold paper before the Royal Society in London, in which he argued that light, like sound, was propagated as a wave. As proof, he devised the double slit diffraction experiment. He argued that the different colored rings produced in the experiment could only occur if light was a wave. He then calculated the size of the wavelengths of the different colors of light. To reinforce his theory, he further noted that two crossed beams of light did not obstruct each other (as expected if a light beam was a stream of particles). Note, however, that Young agreed with Newton's thought that the sensation of color is created inside of us.

The next great young mind to step onto the stage of color vision was James Clerk Maxwell. Maxwell was yet another young Cambridge student, who would become a "Giant of Light." At age 24, he introduced the revolutionary concept of electric and magnetic fields filling all of space. He then hypothesized that when excited, the fields produce photons. Just as an ocean wave comes from the ocean itself, so photons come from an excited electromagnetic field. Space, therefore, was permeated with an active media, perhaps put there as a result of the "original explosion" that started the universe. Frank Wilczek and Betsy Devine[3] suggest that one can think of space as a universal sounding board. The board can be excited at different places. When the board starts to vibrate, the vibrations streak in a wavelike form at a finite speed. But how are electric and magnetic waves and fields related to light? Using measurements from his experiments, Maxwell calculated the speed of electromagnetic waves to be 3×10^{10} cm/sec. But this was the speed of light as recently measured by the French physicist Jean Foucault. Maxwell wrote, "This velocity is so nearly that of light, that it seems we have strong reason to conclude that light itself (including radiant heat and other radiation if any) is an electromagnetic disturbance in the form of waves."[4]

Thus, Maxwell not only gave us the field concept, but by unifying electricity and magnetism with light, led us to the idea that within the cosmos there is a broad electromagnetic spectrum. But what was Maxwell's role in color vision? In 1861 he produced the first color photograph. He followed Young's ideas quite literally to accomplish this feat. He made three black and white photographic slides of a colored plaid ribbon: one taken through a red filter, a second through a green filter, and a third through a blue filter. Then he placed the three slides in three different projectors with the three original filters over each projector. When he pointed all three projectors onto a screen and superimposed the three images onto the same part of the screen a perfect rendition of the colored plaid ribbon appeared.[5]

However, there was something missing. Certain color sensations could not be explained by the trichromatic theory. No mixture of three color filters and projectors could produce brown or olive green. Finally, if one took a brown spot and excluded all surrounding light by looking at the spot through a tube, the brown became a yellowish orange. This result suggested that the surrounding color and the surrounding intensity influences the apparent color of an object. The visual system seemed to be able to create new color effects or alter the perception of certain colors. As you might guess from the material in the last chapter, the brain could be responsible for inhibiting the effects of certain colors while enhancing the effects of others. Interestingly, this line of reasoning was first suggested by Ewald Herring[6] about 100 years ago. More about the brain's influence on color later.

In the 1950s another young scientific genius, Edwin Land, the inventor of instant photography and the founder of the Polaroid Corporation, entered the scene. As opposed to the other three Cambridge graduates, Land was an American who never completed college. Let us start with Land's own words. "I was working on color for one-step photography and I had to have a whole new process. I started from the bottom—that is from the work of James Clerk Maxwell." Recall that Maxwell had produced the first color photograph in 1861 by making three black and white slides of a plaid ribbon using three different color filters and then simultaneously projecting those three black and white slides through three projectors, each covered by one of the original colored filters.

Then, fate and the prepared mind of Edwin Land teamed up to take the theory of color vision to a new level. As Land was repeating Maxwell's experiment, the green filter fell off one of his projectors. "Now it was projecting white light instead of green," Land recalled, "Yet I noticed that the color of the projected image stayed pretty much the same." Then he turned off the projector with the blue filter. "So now I was projecting only the long wave photograph with red light and the medium wave photograph with white light," he said. According to the Young trichromatic theory, the projected slide should have been made up of only red, white, and various shades of pink. Not so! The projected image had all the colors. The colors must have been created by Land's eye and brain. Yes, the brain must play a role in color vision, as Herring's work had suggested. Land went on to observe that a yellow lemon looks yellow whether you illuminate it with the white light of the midday sun or the reddish light of sunset. Ironically, Thomas Young had observed the same thing in 1807 when he wrote, "so when a room is illuminated by the yellow light of a candle or by the red light of a fire, a sheet of writing paper still appears to retain its whiteness." Both scientists had noted than an object remains a constant hue despite the wavelength of light illuminating the object. Although Young did not attempt to explain this color constancy in his color vision theory, Land felt that it was the key element for a new theory. He called his theory the retinex theory, because he assumed color information was processed both in the retina and the cortex of the brain. Land suggested that we do not determine the color of, for example, a red berry in isolation, but always compare the ratio of wavelengths coming from the berry to the ratio of wavelengths coming from the background, like a green leaf. Thus, in the reddish light of late afternoon, the berry will reflect more red light than the green leaf and less of the green. Specifically, the ratio of color readout for the berry will be high red, low green, and lowest in blue. The leaf will read lowest in red, low in blue, and high in green. Let us take a look at the same scene in the greenish light produced when all the light of the sky is filtered through the translucent leaves of a forest. When the light is predominantly green, the red berry will produce a red reading (some red passes through the green leaves), a lower green reading, and lower blue. However, the surrounding green leaf produces a very high green and intermittent blue and a very low red reading. Therefore, the ratios of red versus green from berry and leaf are about the same whether illuminated by red or green light. Land felt that these ratios were computed somewhere in or between retina and cortex.[5]

1. NEURAL PROCESSING OF COLOR

Within the past 20 years, regions in the brain of the monkey and human have been isolated that compare all colors of an object with its surround. These areas are called *blob areas,* because they appear as blobs of cells scattered throughout the visual cortex. In some cases, the cells triggered by the surround will seem to enrich the color of the target, while in other locations, cells representing the surround seem to weaken the color of the target. As noted earlier, such a system, known as an *opponent color system,* was first predicted by Herring and refined by Land. However, the strongest proof of the human brain's involvement in color vision has come mostly from the careful documentation of the responses of neurologic patients.[7–9,23]

Ironically, the very first case history of a patient who had lost her color vision from brain disease (meningo-encephalitis) was reported in 1664 by a nonphysician. Robert Boyle, the brilliant English scientist responsible for such insights as Boyle's gas law,* turned to clinical-scientific issues later in his life, and published a pamphlet entitled "Some Uncommon Observations About Vitiated Sight."[10]

A little over 200 years later (1881), the German ophthalmologist Steffan[9] examined a gentleman who had been a professional color printer and had suffered a stroke. Upon recovery, the printer regained normal visual acuity and normal visual fields but lost his sense of color. Steffan's report suggested a special location in the brain that processed color vision exclusively. But where was this location? A year later, the American neurologist N.E. Brill[11] reported on a patient who had also suffered a stroke that also left him with normal visual acuity and normal visual fields, but with severely impaired color vision. At autopsy, the lesion was noted to be in the rear portion of the visual cortex (posterior part of the gyrus lin-

---

* Boyle's gas law states that the pressure of a given mass of gas is inversely proportional to its volume.

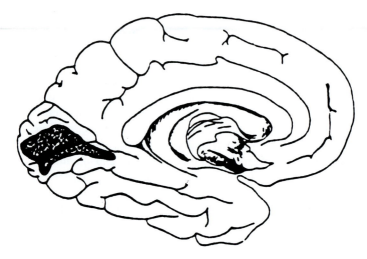

FIGURE 9.2 The lesion in a patient suffering from cortical achromatopsia (From Brill ME: A case of destructive lesion in the cuneus accompanied by color blindness. *Am. J. Neurol. Psych.* 1:356–368, 1882, with permission.)

gualis) (Figure 9.2). What functions did this cortical color vision center actually perform? In 1890, the German neurologist Lissauer[7] described a patient with brain disease, "who could not connect certain color concepts with verbal or complex abstract ideas." This interpretation of the patient's symptoms suggested an integrating role for the color center in the brain. It helped connect color to visual and mental impressions. What exactly does the center do with the color data? We are fortunate that in 1987, two sophisticated clinicians and a neuroscientist had the opportunity to examine extensively a "rather successful artist" who lost his color vision after a car accident. Their report places the work of the cortical color center in very human terms. What follows are some of the artist's own observations.[8]

To him, people now looked "like animated gray statues." He saw "life in a world molded in lead. ... Food looked dead and disgusting." We must remember that this man had seen color in the past. Thus, he was comparing his new world to his old world. Faces were often identifiable only if seen up close. His dreams were "washed out and pale or violent, lacking both color and delicate tonal gradients." His personal impression of color constancy disappeared in that the shades of gray that replaced color changed when the color of the illuminating light was changed. Ultimately, the artist changed his lifestyle and became a "night person." He would only explore new cities and new places at night. In this world, normally without color, he was not at a disadvantage.

In some ways, you might think of the normal brain as using color to create an altered perception of the retinal image, much as the visual illusions discussed in previous chapters. For example, colored objects in partial shadow appear to be of uniform color. Alternatively, certain color contrasts are accentuated, and finally a colored object looks the same regardless of the illuminating wavelength of light. There is also good clinical evidence to support the idea that the "color constancy phenomenon" is processed in the brain. Specifically a patient with lesions of the fusiform and lingual gyri was extensively studied and shown to only lose color discrimination, color naming, and color constancy.[12] Finally, another piece of evidence suggests another advantage for the neural processing of color. In a recent study,[13] subjects were shown 48 scenes (half in color, half in black and white). When shown these scenes again, the subjects remembered the colored scenes better than the black and white ones.

2. INCREASED BRIGHTNESS UNDER YELLOW LIGHT

There is just a certain amount of useful daylight in a 24-hour cycle. Modern humans have stretched the number of hours of useful light by using artificial light. The first form of artificial light was fire light. A wood fire light is primarily yellow-orange in color. Please indulge my need to hypothesize relationships between physiologic and behavioral data for a moment. Wouldn't it be remarkable if we perceived everything to be brighter than it really was in yellow or yellow-orange light: If somehow we could extend daylight by some physiologic amplification system? We do know that if you put on a pair of yellow sunglasses indoors, everything in the room looks brighter. Surprisingly, such an augmented sensation actually defies the laws of physics. A colored filter, no matter the color, must absorb some of the incident light, and therefore transmit less to you, the observer. Yet, as seen in Figure 9.3 (see color insert), the part of the scene covered with the yellow filter looks brighter, as if the yellow light has triggered an amplification device. Such a yellow triggered brightness amplifier would have been helpful to our ancestors. Imagine for a moment the benefit of a perceived brighter light by our primitive ancestors while they were sitting around the yellow light of a fire. With a brighter perceived light, the early cave dwellers could interact more productively indoors. Visual communication could take place in fire light. Such amplification would have made it easier to both produce and appreciate the cave paintings of Cro Magnon man, as found in southern France and northern Spain. In Figure 9.4 (see color insert), we see one of those paintings. Note the use of colors that look brightest in yellow light.

When I suggested this theory to a colleague, she was troubled: Didn't I realize that psychological brightness was different from physical brightness? One needed physical brightness to see indoors, and only a big fire would yield enough light to yield productive light. One cannot increase brightness mentally. Only the intensity of the flame, not its color, was the important variable. I asked her to put on a pair of yellow spectacles to actually experience the phenomenon. "No, that won't be necessary, a colored filter can only diminish light not amplify it," she said.

Indeed, she had not yet learned that there is no such thing as a ball or a strike until the umpire calls it.

3. COLOR CONSTANCY

Earlier in the chapter, it was noted that a red berry against a green leaf looks the same no matter the time of day. To the physicist, such an observation is surprising because the light striking the earth from our sun changes its mixture of wavelengths throughout the day. Of course, it is not the light itself that changes but the properties of the atmosphere that absorb and scatter different wavelengths differently during the day. Such changes account for the blue of the sky at midday and the reds and violets at day's end. Figure 9.5 (see color insert), is a series of photographs that dramatize the color shifts in the sky throughout the day. Recall that Edwin Land[5] put forth his retinex theory in which he credited the retinex for maintaining the perception of constant colors as long as the ratio of wavelengths reflected from the berry and leaf are the same. Clearly, such a constancy preserves the set of clues needed by a primate to distinguish ripe, sweet berries from the unripe, bitter variety, no matter the time of day.

4. ENHANCEMENT OF COLOR CONTRAST:

There are occasions when the same color will look different when set against backgrounds of different colors (the reverse of color constancy). For example, in Figure 9.6 (see color insert), the four colored disks are the same on both the right and left sides of the figure. Although only the background colors differ, the disks themselves appear quite different on either side. Note how orange and red disks appear lighter against the blue background. In a word, the contrast is enhanced.* Not only do the colors of disks look different, but so do the borders of the disks. Note that the disks aligned on the blue background almost appear to have been outlined in pencil. This sharpening of the border stands in direct contradistinction to the expected smear that you would expect from the chromatic aberration of the eye. You might consider these as colored versions of the illusions described in the preceding chapter.[14] When might such an alteration in the perceived color be of use to us? It is well recognized in nature that a creature (snake, insect) that has a skin pattern containing two or more bright colors usually has glands that produce an unpleasant or poisonous chemical. The poisonous creature or its mimic uses its bright-colored combination as a warning to stay away. For example, in Figure 9.7 (see color insert), we see a picture of the toxic oil beetle, Mylabris oculata, (also

---

* A colored object against a colored background is not always enhanced. Firefighters note that they cannot see the yellow flames of a forest fire against a light blue sky. They follow the fires progress by watching the black smoke against the light sky.

FIGURE 9.3 Note how the portion of the scene covered with a yellowish filter looks brighter, yet receives less light because of the light absorption properties of the filter.

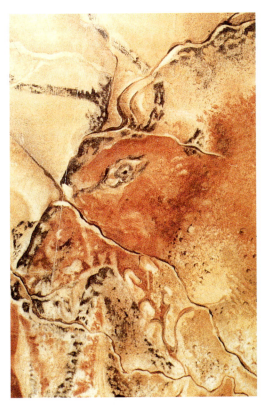

FIGURE 9.4 A cave painting done about 25,000 years on the wall of a cave in southern Europe. Note the use of yellow and brown, both accentuated in yellow light. (From the cover of *The American Scientist,* Dec. 1993, with permission.)

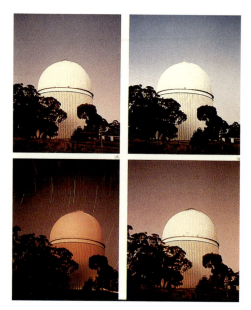

FIGURE 9.5   Four photographs of an observatory taken at four different periods during the day. Note how the color of the sky changes throughout the day. (From Malin D, Murdin P: *The Colours of the Stars*. Cambridge University Press, Cambridge, 1984, p. 77, with permission.)

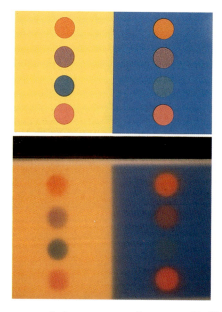

FIGURE 9.6   The phenomenon of color opponency, as first expressed by Herring, is represented in this picture. Note how the four colored disks against a yellow background look darker than the same disks against the blue background. Also note the appearance of the outline around the borders of the disks against this blue background. The phenomenon is diminished if the disks are blurred. (Courtesy of Hurvich LM: *Color Vision*. Sinauer Associates, Inc., Sunderland, MA, plate 13–1, 1981, with permission.)

FIGURE 9.7 The toxic oil beetle warns any intruder of its presence with its distinctive coloring. Our enhanced color contrast allows us to see the creature even in a dusty or foggy environment. (From O'Toole C (ed.): *The Encyclopedia of Insects.* Facts on File, New York, 1986, p. 69, with permission.)

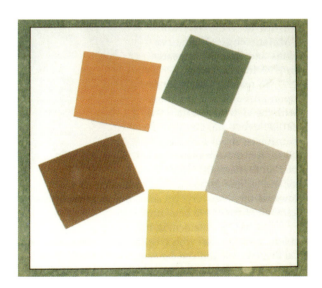

FIGURE 9.8 All the colored squares are perceived as green by different people with different types of color deficiency. (From Rosenthal D: Color confusion. *Optics and Photonics News,* Vol. 5, No. 7, 1994, p. 9, with permission.)

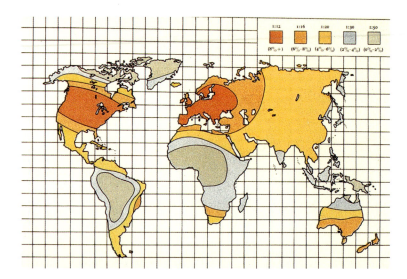

FIGURE 9.9   World map showing the different incidences of color deficiency in different parts of the world. Note that the incidence is 1 in 12 in Europe and the United States, whereas it is very rare (1 in 50) in the more underdeveloped (hunting and fishing areas) portions of the world (From Knapp VH (ed.): *Colour.* Knapp Press, Los Angeles, Viking Press distributors, New York, 1980, p. 37, with permission.)

known as Spanish fly) mating on a leaf. Note the vivid blotches of orange and yellow in its back. This insect contains high amounts of the irritating chemical cantharidin. Any brain-enhancing mechanism that brings greater attention to such a vividly colored creature might be thought of as having a survival benefit for us.

5. COLOR BLINDNESS

Color blindness is a broad category term that includes three major varieties of people who have trouble distinguishing different colors. Aside from those rare cases of brain damage discussed earlier, the deficiencies are usually caused by abnormal pigments in the cones of the retina.* The largest subset (about 75% of those with color blindness) only have difficulty making fine distinctions between colors. For example, all the squares in Figure 9.8 (see color insert), are perceived as green by people with different types of this major type of color deficiency. These people are called color anomalous. A group about one-third the size of the color anomalous subset cannot recognize certain important colors at all and are called dichromats.† Finally, there is a tiny group who see the world only in shades of gray. These people are known as monochromats.[15] For the moment, let's focus on the larger color anomalous group. Surprisingly, they are by and large all male. Even more interesting is the fact that the incidence of color anomalous vision among males varies in different parts of the world as depicted in Figure 9.9 (see color insert). In the United States and much of Europe, almost 1 male in 12 has the condition. In parts of the world where the societies are basically made up of hunters, fishermen, and farmers, the incidence of color deficiency is very low (1 in 50 or less). We do know that hunters must have excellent red-green discrimination when they track wounded animals (they must be able to identify dried blood on dirt, grass, and leaf surfaces). This point was vividly brought home a number of years ago when I visited a private game park in Africa. One of the members of our group, an army colonel, was an avid hunter, and had brought his favorite rifle. When he spotted a wildebeest about 100 yards away, he fired. Unfortunately, he only wounded the animal. Our guide told us that it was our obligation to find the animal and put it out of its misery. Over the next 3 hours, the elderly guide tracked the animal. The trail was composed of a series of dark green leaves covered with subtle spots of dried dark red blood. Within 3 hours, he located the wounded animal, and the colonel killed it.

Is it legitimate to ask if there might be an advantage in being color anomalous in certain settings? Note that the highest incidence of color blindness occurs in

---

* Incidentally, the rareness of color blindness caused by a brain lesion suggests that this part of the brain is very well protected from disease processes and injury.

† People with normal color vision can discriminate about 100 different hues created by mixing three primary colors. Dichromats can identify a much lower number (i.e., these created by mixing two primary colors).

the United States and Europe, areas that were considered very cold climates in the ice age. Hunters living in this part of the world would spend a great deal of time in caves or dwellings sitting around a fire. Let me suggest that it might have been advantageous to have color sensitivities more suited to such an environment. Instead of having the three "normal" cone pigments with peak sensitivities to blue, green, and red, Professor J. Mollon[16,24] of Cambridge University suggests that these men may have had cone pigments with peak sensitivities to blue, red, *or* green and a third anomalous pigment most sensitive to a yellow-orange color or a color between green and red. Since this anomalous pigment is in between red and green in the spectrum, these people would do poorly discriminating between red and green. However, they might be very sensitive to the different shades in the orange-yellow portion of the spectrum. Indeed, in the fire light of those ice age caves, where all the colors are influenced by the yellow-orange light of the wood fire, they might appreciate subtleties in color while painting or weaving that the normal sighted would miss.

A few years ago, a student and I were actually able to lend a bit of support to Professor Mollon's theory. By chance, we had discovered an amazing variable color filter produced by a photographic supply company. By rotating the housing of the filter, the color slowly changed from red through yellow and onto various shades of green. We advertised for color anomalous men and tested their color vision with and without this filter. Interestingly, they performed normally in the orange-yellow region. However, the green anomalous subjects could be helped to see targets, normally invisible to them, by looking through different parts of the green end of the variable filter, and red anomalous men would see targets, normally invisible to them, by looking through the reddish portion of the variable filter. Thus, their own anomalous pigment functioned normally in one region of the spectrum, and benefited from a boost by the addition of more color in their defective color vision regions.[25]

6. EVOLUTION OF COLOR VISION

Color vision developed as the ancient pigment rhodopsin evolved into variants sensitive to red, green, blue, and ultraviolet. In an earlier chapter, it was noted that an ancient bacteria that may have evolved 1.3 billion years ago had the "grandmother" light sensing pigment, bacterial rhodopsin. As insects and plants evolved, it is suggested that a symbiotic relationship between plants and insects developed in which plants exchange food to the insects in return for pollination services. Since color signaling by plants was essential in this relationship, the successful insects evolved different color-sensing pigments. I am suggesting that identifying different flower species, as well as the color changes announcing high nectar content (which can be associated with peak time for pollination), forced the insects to develop very sensitive color vision. Interestingly, insects, as opposed to humans, respond to conventional colors as well as ultraviolet light.[17]

Evolutionarily speaking, excellent color vision also developed in fish and birds. Ironically, reptiles and amphibia developed only poor to fair color vision, as was the case with most mammals.[18] Many mammals looking at a bed of pansies might see them only in shades of black, white, and gray, as in Figure 9.10. Without color, these flowers take on the appearance of faces, which would tend to frighten flower-eating mammals away, and may well be the flower's way of protecting itself against "color blind" predators.

What might some of the factors have been that were responsible for the poor color vision in mammals? One possible explanation is that during the era of dinosaur pre-eminence, mammals were thought to spend most of their days hiding from these large creatures, and only came out at night. The pigments of the cone color vision system function best in bright illumination. At night, cone vision is less active, and most objects are recorded by rods in shades of gray. Therefore, the mammals had little need for a color vision system. If the Alvarez[26] theory is correct, a meteor implosion into the earth lifted huge amounts of dirt and rock into the atmosphere some 75 million years ago, preventing the warmth and light of the sun from reaching the earth. This "long, dark, winter" is thought to be responsible for the extinction of the dinosaur. However, the warm-blooded mammals with their poor color vision did survive in the dim light of the chilling new world.

FIGURE 9.10   Black and white photo of a bed of pansies. In their natural color, this bed of flowers would not resemble faces. To mammals with poor color vision, this facial mimicry may have a protective effect for the flowers.

With the ultimate lifting of the implosion-created cover of darkness, the newly arrived primates were able to survive by collecting ripe fruits and vegetables. It has been suggested,[19] that these feats required the evolution of a new color vision system. This supposition has been brought into question by the findings of Oliver Sacks. In his investigation of totally color blind people on the Pacific atoll of Pinguelap, he noted that these people used touch and smell to determine ripeness.[27] However, the new primate color vision system added something new besides the photopigments found in older species, that is, a major brain processing component.

7. OTHER ADVANTAGES OF BRAIN PROCESSING OF COLOR VISION

What might be the advantages of the "new color vision system"? Recall the "rather successful artist" discussed earlier, who lost his color vision after an accident produced a lesion in his brain. Author, neurologist Oliver Sacks, in his book, *An Anthropologist on Mars,*[20] discusses this very same patient and offers new insight to our question. Although the artist reported that his vision had become much sharper, he could not identify faces until they were close.* Dr. Sacks also tells us that the artist tended to misinterpret shadows. Since many facial expressions and details are related to facial shadows, which change under different lighting situations, we can start to understand how brain processed color vision aids facial recognition. For example, among Caucasians, research on infant vision teaches us that although newborns have poor color vision, they can distinguish shades of yellow and orange from white light[21] (i.e., facial colors). (Incidentally by age 3 months, infants have developed good color vision.)[21] Anthropologist Peter Frost informed me that his research showed that 2- or 3-year-olds were highly sensitive to barely perceptible differences in skin color (using painted dolls) (Personal communication, 1989). Thus, color vision is useful in helping to recognize faces, and it appears at a very young age.

Earlier in the chapter, it was noted that the yellow light of the fire appears to make objects look brighter. It was then suggested that this "internal amplification of brightness" might have helped our early ancestors to paint by firelight as well as communicate (i.e., read facial expressions). All of the above further support the concept that color vision helps facial communication.

Dr. Sacks' artist patient also complained that his dreams as well as his migraine aura were without color. Earlier it was noted that a color picture was remembered more readily than one of black and white.[21]

In sum, the "new" primate color vision facilitates dreams, memory, and the recognition of faces and facial expression (particularly under artificial light), as well as color contrast enhancement.

---

* Patients with achromatopsia because of a retinal lesion do not report many of the problems reported by those whose loss of color vision was caused by a brain lesion.

REFERENCES

1. Zukav G: *The Dancing Wu Li Masters.* William Morrow and Company, New York, 1979.

2. Tyndall J: *New Fragments.* Longmans Green and Company, London, 1882.

3. Wilczek F, Devine B: *Longing for the Harmonies.* W.W. Norton and Company, New York, 1988.

4. Tolstoy I: *James Clerk Maxwell.* University of Chicago Press, Chicago, 1981.

5. Land EH: Recent advances in retinex theory and some implications for cortical computations: Colour vision and natural image. *Proc. Natl. Acad. Sci. (USA)* 80:5163–5169, 1983.

6. Hubel DH: *Eye, Brain and Vision.* Scientific American Library, A division of HPHLP, New York, 1988.

7. Lissauer H: Ein Fall von Seelenblindheit nebst einem Beitrage zur Theorie der selben. *Arch. Psychiat. Nervenkr.* 21:222–270, 1890.

8. Sacks O, Wasserman RI, Zeki S, Seigel RM: Sudden color blindness of cerebral origin. *Soc. Neurosci. Abst.* 14:1251, 1988.

9. Steffan PH: Beitrag zur Pathologie des Farbensinnes. *Graef. Arch. Ophthalmol.* 27:1–24, 1881.

10. Boyle R: Some uncommon observations about vitiated sight. In *Robert Boyle, The Works* (Birch T, ed.). Vol. V reprint 1965, Olmes, Hildeschein, London, 1664, pp. 445–452.

11. Brill NE: A case of destructive lesion in the cuneus accompanied by color blindness. *Am. J. Neurol. Psychiat.* 1:356–368, 1882.

12. Morland A, Ruddock KH: Color constancy in a patient with impaired color vision associated with lesions of the fusiform and lingual gyri. *Invest. Ophthalmol. Vis. Sci.,* Abstract Book 36, No. 4, 1995, p. 5470.

13. Wichman FA, Gegenfurther KR: Color recognition and image understanding. *Invest. Ophthalmol. Vis. Sci.,* Abstract Book 36, No. 4, 1995, p. 514.

14. Robinson JO: *The Psychology of Visual Illusion.* Hutchinson, London, 1972.

15. Hurvich LM: *Color Vision.* Sinauer Associates, Sunderland, MA, 1981.

16. Mollon JD: Color vision and color blindness. In *The Senses* (Barlow HB, Moullon JD, eds.). Cambridge University Press, Cambridge, 1982.

17. Menzel, R: Spectral sensitivity and color vision in invertebrates. In *Handbook of Sensory Physiology,* Vol. VII/6A: Invertebrate Visual Centers and Behavior. I (Autrum H., ed.). Springer, Berlin, 1979, pp. 503–580.

18. Walls GL: *The Vertebrate Eye and Its Adaptive Radiations.* Reprinted by Hafner Publishing, New York, 1963, pp. 663–691.

19. Mollon JD: The uses and origins of primate color vision. *J. Exp. Biol.* 146:21, 1989.

20. Sacks O: *An Anthropologist on Mars.* A.A. Knopf, New York, 1995.

21. Brown AM: Development of visual sensitivity to light and color vision in human infants: A critical review. *Vision Res.* 30: 1159–1188, 1990.

22. Gregory RL: *The Intelligent Eye.* Weidenfeld and Nicholson, London, 1970.

23. Zeki S: The representation of colours in the cerebral cortex. *Nature (London)* 284:412–418, 1980.

24. Mollon JD: The Tricks of Color. In *Image and Understanding* (Barlow HB, Blakemore C, Weston-Smith M, eds.). Cambridge University Press, Cambridge, 1990.

25. Bearse M, Miller D: Improved color vision testing. *Ann. Ophthalmol.* 14(2):344, 1982.

26. Alvarez W, Asaro F: An extra terrestrial impact. *Scientific American,* Oct. 1990, pp. 78–84.

27. Sacks O: *The Island of the Colorblind.* A.A. Knopf, New York, 1997.

10

AWARENESS OF THE RETINAL IMAGE

1. Visual Self-Awareness
2. Visual Awareness
- Summary

1. VISUAL SELF-AWARENESS

Aside from the human being, only chimpanzees recognize themselves in a mirror. However, chimps in captivity reared in isolation do not recognize themselves in the mirror.[1] From such experiments we might surmise that visual self-awareness is only useful for primates in a social setting.

We not only recognize ourselves in the mirror, but we also recognize our own printed name 10 times more quickly than the names of others.[2]

Why is it so important that we are aware of our own images as well as images of our names? We might get a clue to the answer by observing a patient with autism. In a book entitled *Somebody, Somewhere,* author Donna Williams,[3] a very intelligent and articulate person who suffers from autism, tells us that her "self" was nowhere to be found. She had no self-image of who she was. For example, when she looked at herself in the mirror, she thought that the person she saw in the mirror was someone else who kept her company. She sometimes suggests that she is a nonentity in a friendless world. She cannot form a sincere relationship with anyone. She writes, "I was allergic to words like `us' or `together' ... words depicting closeness made earthquakes go off inside me and compelled me to run." Ms. Williams' words suggest that without an awareness of herself, a recognition that she actually exists and has a special place in the world, she cannot make sincere social relationships. I use the words "sincere relationship" because she tells us that she had learned to make small talk in order to pretend at relating to others. Perhaps sense of self (visual and otherwise) is the reference point from which we launch our connections with others.

If we can speculate on Ms. Williams' own words, it would appear that we use an awareness of our own visual image when we deal with others. Such awareness may help us to quantitate certain characteristics of others such as tallness, thinness, beauty, and so on, by using our own appearance as a comparator.

Of course, our image is continually changing, be it from aging, accidents, diseases, or experience. Therefore, our sense of what we look like is continually edited and updated. We have no data concerning young children living in isolation, but we do know that a child normally recognizes itself in a mirror by age 2.[4] Thus, in the human, a visual sense of self comes quite early in development, suggesting that it might be fundamental to our socialization.

2. VISUAL AWARENESS

Patients with two unusual brain conditions have symptoms that support the notion that a visual awareness center exists.[5] The first condition, known as "denial of blindness" or Anton's syndrome, is usually caused by a tumor or injury of the brain, or a cerebral vascular stroke that affects both sides of the visual cortex.[6] Patients who are afflicted with Anton's syndrome are usually totally blind. Yet, their behavior is very different from a typically blind person. If you were to place an object in front of such a patient and ask them if they can see it, they will answer in the affirmative. If you then ask them to describe the object, they will invent an answer. When they are told that the answer is wrong, they will invent an excuse. For example, suppose that you place a pencil on a desk in front of a blind patient with Anton's syndrome and ask her to describe the object. The patient might suggest that the object is a piece of paper. When you tell the patient that the object is a pencil, the response may be something like, "Oh, well of course I saw the pencil, but I didn't think that was important enough to mention." The visual awareness center is probably working in these patients. However, they are responding to imagined images instead of the real ones, and they do not seem to appreciate the difference.

There is a second neurologic syndrome that adds further support to the argument that there exists a center of visual awareness. It is a condition known as *blind sight.*[5-7] These patients have also suffered severe damage to their visual (occipital) cortex. Using conventional visual field testing, they too are either totally blind or blind in a sizable part of their visual field. If the examiner places an object within their blind area, their response is diametrically different from the patient with Anton's syndrome. They will tell you that they cannot see the object, the expected response of a blind person. However, supposing you say, "I know that you cannot see this object, but simply try to guess where it is and touch it." Remarkably, they will accurately touch the object almost every time. These patients can see, but their visual awareness center has been destroyed. They do not know, on a conscious level, that they are seeing.

FIGURE 10.1 A cave painting at Cosquer, France. From paint samples, experts concluded that the painting was done 30,000 years ago, making it the oldest cave painting ever found. (From *Time Magazine,* June 19, 1995, p. 49, with permission.)

There even seems to be a center for color vision awareness versus black and white vision. A subject with cortical blindness reported that he could only see black and white. Yet, when shown different colored squares of the same light intensity that touched each other, he was able to differentiate them.[8]

SUMMARY

You might think of the motto of the visual awareness center as, "you shall bear witness." Bearing witness to an event implies that you are able to ultimately tell someone else about an event that you observed. As social creatures, we continually tell each other what we have seen. I am suggesting that if we simply responded to visual events in an unconscious fashion (i.e., as patients with blind sight) we could not share the event with others. Perhaps an anecdote will help.[7] Bill Bradley, a famous basketball player (later a U.S. Senator), was practicing in a gym he had never played in before. For the first few minutes of practice, he missed every shot he attempted. Then he stopped, seemed to make a mental adjustment and then made four perfect shots in a row. Then he paused and said, "You want to know something, that basket is about an inch and a half low." A measurement showed the basket was an inch and an eighth low. Bringing his personal observation to the awareness level was useful to others, as well as himself.

A second advantage of a visual awareness center relates to our ability to not only talk about events but to draw events. Figure 10.1 (see color insert), is a photo of a drawing on the wall of a cave at Cosquer (near Marseille). Samples from the paint date the time of the painting to 30,000 years ago, making it the oldest cave painting ever found. Observers have noted that the artist used the natural contours of the wall to give the work almost a three-dimensional effect. Obviously, the artist could not bring the animal inside the cave and freeze it as a model. The artist used his or her visual memory and visual awareness to recreate the personal impression of the creature.

Perhaps this whole chapter might be summarized in the words of the philosopher Emanual Kant,[2] when he wrote "Sensory input without abstract labeling equals blindness." Must we bring our observations into the visual awareness center, in order to create abstractions of these observations?

REFERENCES

1. Gattup GG Jr, McLure MK, Hill SD, Bundy RAC: Capacity for self-recognition in differently reared chimpanzees. *Psych. Rec.* 21:69, 1971.
2. Mack A, Hartman T: Your name pops out. *Invest Ophthalmol. Vis. Sci.* Abstract Book 36, no. 4, 1995, p. 5902.
3. Williams D: *Somebody, Somewhere.* Time Books/Random House, New York, 1994.
4. Landau T: *About Faces.* Anchor Books, New York, 1989.

5. Grusser O, Landis T: Blindsight In *Visual Agnosias and Other Disturbances of Visual Perception & Recognition* (Grosser D, Landis T, eds.). CRC Press, Boca Raton, FL, 1991, p. 148.

6. Weiskranz L: *Blind Sight: A Case Study and Implications.* Oxford University Press, Oxford, 1986.

7. Jayawardhana R: Unravelling the Dark Paradox of Blind Sight. *Science* 258:1438, Nov. 27, 1992.

8. Heywood CA, Cowley A, Newcombe F: Chromatic discrimination in a cortically blind observer. *Eur. J. Neurosci.* 3:802, 1995.

5. Jackendoff R: *Consciousness and the Computational Mind.* Bradford Books, MIT Press, Cambridge, MA, 1987.

6. Johnson-Laird PN: *The Computer and the Mind: An Introduction to Cognitive Science.* Harvard University Press, Cambridge, MA, 1988.

11. McPhee J: *The Ransom of Russian Art.* Farrar, Straus & Giroux, New York, 1995.

12. (Quoted in) Zajonc A: *Watching the Light.* Bantam Books, New York, 1993.

INDEX

Abrasion, corneal, 62–63
Accommodation, 32–33
Adrenaline levels, pupillary response and, 86–88
Age
  eye appearance and, 92
  visual function and, 41, 142–144
Air or land and water niche, 57–58
Alternating squint, 15
Alvarez theory, 157
Ambiguous figures, 123–124
Amblyopia, 15–16, 22–23
Anthropologist on Mars, An, 158
Anton's syndrome, 162
Apparent motion, 124–127
Aspheric surfaces, 39
Astigmatism
  cultural considerations, 78–79
  in infants, 10–11, 13
  optical considerations, 77
  related factors, 77–78
Axial length, 4–8, 52

Baron-Cohen, Simon, 13, 90
Belkin, Michael, 61
Benedek, George, 29
Biologic attraction, eye contact and, 89
Birds, diving, 57–58
Black sclera
  of eagles, 47
  of monkeys, 49

Blind sight, 162
Blind spot, 129
Blink rate, 90–91
Blob areas, 151
Blue light, light scattering and, 37
Bodamer, Joachin, 131
Boyle, Robert, 151
Brain
  contribution to vision, 104–105
  processing in animals, 106–107
  processing of color, 151–153, 158
  retinal image enhancement and, 135–142
  contrast enhancement, 138–140
  edge sharpening, 140
  filling in information, 135–138
  removing distractions, 142
  vernier acuity, 141–142
  as story teller, 105–106
Brill, N.E., 151
Brow
  light scattering and, 38
  positioning of, 91–92
Bruce, Vicki, 131
Butterfly eye spots, 94–95

Cataracts
  from blunt trauma, 66–68
  in later childhood, 15–16
  traumatic in child, 65–66
  treatment for, 31

Cats, 58
Chaos and Order, 104
Child art, 16–21, 23
    culture and, 19–21
    drawing faces, 17–19
Children. *See* Child art; Infants and children
Chromatic aberrations, 38–39
Cohn, Herman, 74
Color anomalous, 155–156
Color blindness, 155–156
Color constancy, 154
Color contrast, 154–155
Color vision
    brain processing advantages, 158
    brightness under yellow light, 153–154
    color blindness, 155–156
    color constancy, 154
    color contrast enhancement, 154–155
    evolution of, 156–158
    historical overview, 147–151
    neural processing of color, 151–153
Compound eye, in insects, 54
Contrast enhancement, 138–140
Cormorants, 57–58
Cornea
    abrasion of, 62–63
    of diving birds, 57–58
    of eagle, 46
    of fish, 56–57
    lacerations of, 63–66
    light scattering and, 37
    role of, 28–2
    scratched, 63
    spherical aberration and, 39–40
Craik-Cornsweet-O'Brien illusion, 140
Cramer, Friedrich, 104, 109–110
Crying, 88
Crystalline lens, role of, 30–31
Crystallins, 31
Culture, child art and, 19–21

Denial of blindness, 162
Deprivation amblyopia, 15–16
Destructive interference, 37
Devine, Betsy, 149
Devore, Irvin, 89
Diffraction patterns, 35–36
    of insects, 54–55
Dilation, pupil, 86–88
Disappearing images, 127–128

Distant objects and illusion, 115–118
Drawing faces, 17–19
Drifts, 36
Driving, visual acuity and, 143–144

Eagles, 46–49, 58–59
Edelhauser, Hank, 56
Edge sharpening, 140
Ehrenstein figure, 17
Einstein, Albert, 34
Emmetropization, 7–8
Enoch, Jay, 38
Evolution
    of color vision, 156–158
    of ocular components, 41–42
Eye communication
    advantages of, 92
    animal eye patterns, 97
    animal eye spots
    butterfly eye spots, 94–95
    eye spots in other species, 95–97
    determining health and, 93
    human
    blink rate, 90–91
    crying, 88
    determining age, 92
    eye reddening, 88–89
    eye to eye contact, 89–90
    positioning of brows and lids, 91–92
    pupillary response, 86–88
    man-made eye patterns, 98
    in painting, 93
Eye contact, 89–90
Eye injuries. *See also* Eye repair
    blowout orbital fracture, 68–70
    blunt trauma, 66–68
    corneal abrasion, 62–63
    corneal laceration, 63–66
    iridotomy, 66–68
    occurrences per year, 61–62
    traumatic cataract, 65–68
Eye lens, spherical aberration and, 40
Eyelids, light scattering and, 38
Eye movements, infants monitoring of,
    13–14
Eye reddening, 88–89
Eye repair. *See also* Eye injuries
    in animals, 70–71
    natural, 70
Eye size
    myopia and, 76
    in tall animals, 52

Eye spots
  butterfly, 94–95
  in fresh water fish, 96
  in frogs, 95
  in ocean fish, 95
  in peacocks, 96
  in reptiles, 95

Facial expression, 91–92
Facial recognition, 131–132
  color vision and, 158
  infants, 9–12
Farsightedness. *See* Hyperopia
Field of vision, 40
Filling in information, 135–138
Flashing lights phenomenon, 127
Fovea, 8
Foveal reflex, 14

Gap figures, 137
Gap illusion, 137
Gene sharing, 42
"Getting around" vision, 70
Giraffes, 51–53
Glare, 53
Global impressions, 128–129
  vs. detail detection, 130–131
Glycosaminoglycan, 28–29
Goffman, E., 89
Goldman, Jerry, 29
Gregory, Richard L., 16

Health, determining through eyes, 93
Hermann, Ludimar, 124
Hermann grid, 125
Herring, Ewald, 150
Hess, Ekhard, 87
High-altitutde niche, 46–49
High off the ground niche, 51–53
Hitting blind, 119
Holobacterium halobium, 34
Horton, Jonathan, 104
Hubel, David, 13
Human embryo, cross section of head, 104
Human faces, registering, 108–113
Human ocular optics
  field of vision, 40

optical aberrations, 36–40
visible light waves, 28–36
Hyperopia
  cultural considerations, 80
  optical considerations, 79
  related factors, 80

Infants and children
  blockage of visual development, 15–16
  child art and visual development, 16–21
  corneal laceration in, 65–66
  early anatomy, 4–9
  early physiology, 9
  hyperopia in, 79
  line orientation receptors, 13
  monitoring of eye movements, 13–14
  recognition of three dimensions, 14
  recognizing faces, 9–12
  recognizing movement, 14
  traumatic cataract in, 65–66
Information processing, in aging, 143
Insects, 53–55
Iridotomy, from blunt trauma, 66–68
Iris, prolapsed, 64

Kanisza figure, 137–138
Keane, Walter, 93
Kellog, Rhoda, 17

Lacerations, corneal, 63–66
  in children, 65–66
Land, Edwin, 150–151
Landolt rings, 138
Lens proteins, 65
Lettvin, Jerry, 106
Lid positioning, 91–92
Light scattering, 36–38
  natural defenses against, 37–38
Light waves, 28–36
Line orientation receptors, 13

Mammals, color vision in, 157
Mandela pattern, 17
Marr, David, 104
Master regulatory gene, 59
Maxwell, James Clerk, 149, 150
Microsaccades, 36
Mindblindness, 13

Mollon, J., 156
Monkeys, 49–50
Monochromats, 155
Moon illusion, 118–119
Mountcastle, Vernon B., 105
Movement, infant recognition of, 14
Muller-Lyer illusion, 114
    horizontal, 120
Myopia
    cultural considerations, 75–77
    optical considerations, 74
    related factors, 74–75

Nakanishi, Koji, 34
Nerve cell loss, in aging, 142
Neural processing, 8–9
    color, 151–153
Newton, Sir Isaac, 147–149
Nighttime niche, 58

Ocean niche, 56–57
Ocular components, evolution of, 41–42
Odd Perceptions, 16
"Old age sight." See Presbyopia
Opponent color system, 151
Optical performance, extremes of, 58–59
Orbital fracture, blowout, 68–70

Painting, eye communication in, 93
Pecten, 47
Peli, Eli, 108
Penguins, 58
Photoelectric effect, 34
Photoreceptors, retinal, light scattering and, 38
Physiology, early, 9
Presbyopia, 33
    cultural considerations, 82
    optical considerations, 80–81
    related factors, 82
Proptosis, 87
Pterygium, 77
Pupillary response, 86–88
Pupil size, 47

Radiating patterns, 126
Ramachandran, V.S., 129
Receptor size, 35–36
Red light, light scattering and, 37

Refractive error
    astigmatism, 77–79
    hyperopia, 79–80
    myopia, 74–77
    presbyopia, 80–82
    society and, 83
Removing distractions, 142
Retina
    of eagle, 49
    light scattering and, 37–38
    role of, 33–36
Retinal image, 28
    brain as story teller, 105–106
    brain mechanisms enhancing, 135–142
    brain processing in animals, 106–107
    color vision and. See Color vision
    history of brain processing, 104–105
    visual awareness center, 162–163
    visual illusions. See Visual illusions
    visual self-awareness, 161–162
Retinal magnification factor, 51
Retinal receptors, 8
Rhodopsin, 34, 156
Rotation simulation, 125–126

Sacks, Oliver, 106, 158
Santa down the chimney illusion, 128
Scheiner, Father Christof, 104
Scratched cornea, of sculpins, 56–57
Scratches, corneal, 63
Sculpin, 56–57
Self-awareness, visual, 161–162
Size constancy, 117
Slatt, Bernard, 119
Social seeing, 21–22
Somebody, Somewhere, 161
Spherical aberration, 39–40
Stabilized image, 128
Stegmann, Robert, 46, 63
Stein, Harold, 119
Stereopsis, 122–123
Stereoscopic vision, of monkeys, 50
Stern, John, 90
Steroscopic illusions, 120–123
Surgery, for corneal lacerations, 63–65
Sympathetic ophthalmia, 61

Tall animals, 51–53
Tapedum lucidum, 58
Three dimensions, infants recognition of, 14

Time compensating illusions, 119–120
Trauma. *See* Eye injuries
Treatsie on Physiologic Optics, 36
Tree niche, 49–50
Tremors, 36
Turnbull, Colin, 117
Tyler, Christopher, 93

Valsalva maneuver, 89
Variable illusion, 123–124
Vernier acuity, 141–142
Vertical illusions, 113–115
Very close to the ground niche, 53–55
Visual awareness center, 162–163
Visual development, blockage of, 15–16
Visual illusions
    ambiguous figures, 123–124
    apparent motion, 124–127
    background of, 107–108
    disappearing images, 127–128
    emphasizing important objects, 113–119
    global impressions, 128–129
    registering human faces, 108–113
    stereoscopic illusions, 120–123

subjective aspects of, 129–130
    time compensating illusions, 119–120
Visual resolution, of monkeys, 50
Visual surprise, 108
von Helmholtz, Herman, 36

Wade, Nicholas, 132
Walsh, David, 143
Weisel, Torsten, 13
Weiter, John, 41
Wilczek, Frank, 149
Williams, Donna, 161

Yellow light, increased brightness under,
    153–154
Yellow pigment, light scattering and, 37
Young
    Andrew, 131
    Thomas, 148–149, 150

Zukav, Gary, 148